EARTHQUAKE RESISTANT BUILDING CONSTRUCTION

NEELAM SHARMA

EARTHQUAKE RESISTANT BUILDING CONSTRUCTION

By
NEELAM SHARMA
M.E. (Structures with Distinction,
Delhi College of Engineering)

S.K. KATARIA & SONS®
Publisher of Engineering Books
4885/109, Prakash Mahal, Dr. Subhash Bhargav Lane,
Opposite Delhi Medical Association, Daryaganj, New Delhi–110002 (INDIA)
Phone: +91-11-43551243; Mobile: +91-9871775858, 7678425209
e-mail: katariabook@gmail.com, katariabooks@yahoo.com
Website: www.skkatariaandsons.com
(e-books, subscriptions and bulk buying also available on our website)

Published by:

S.K. Kataria & Sons®
4885/109, Prakash Mahal, Dr. Subhash Bhargav Lane,
Opposite Delhi Medical Association, Daryaganj, New Delhi–110002
Phone: +91-11-43551243, Mobile: +91-9871775858, 7678425209
e-mail: katariabook@gmail.com, katariabooks@yahoo.com

First Edition: 2007; Fourth Edition: 2018; Reprint: 2026

Laser Typesetting: Plus Computers, Delhi *Printed at :* New A.S. Printing Press, Delhi

Preface to the Fourth Edition

It is indeed a great pleasure to present the edition of this comprehensive books. The subject of earthquake resistant building construction is evolving daily in the sense that more and more seismic data is available everyday on account of various earthquakes occurring worldwide. As a result of this additional data and further knowledge and experience gained. Bureau of Indian Standards have revised the codes related to Earthquake Resistant Design of Structures recently. To incorporate the provisions of these latest Indian Standards, this edition of the book is very much needed. The present book focuses on the following latest revised codes.

IS 1893 (Part 1): 2016, Indian Standard for Earthquake Resistant Design of Structures.

IS 4326 : 2013, Indian Standard for Earthquake Resistant Design and Construction of Buildings.

IS 13920 : 2016, Indian Standard for Ductile Detailing of Reinforced Concrete Structures Subjected to Seismic Forces.

IS 13935 : 2005 Indian Standard Guidelines for Repair and Seismic Strengthening of Buildings.

In addition to the above, substantial new matter in the form of text, examples and drawings are added in order to understand the codes provisions in a simple way. The book also contains a number of model question papers and examination papers which will be of immense help for the students.

I take this opportunity to express my sincere thanks to all my friends who were good enough to send me suggestions for this book.

Any comments/suggestions to improve the utility of the book will be gratefully received.

February, 2018 **Neelam Sharma**

Preface to the First Edition

Earthquakes are one of the most destructive natural hazards. They not only cause damage to the property but are responsible for a large number of lives. The loss of lives is mainly because of collapse of structures. Till to date, man is not able to predict and prevent this dangerous event, caused due to shaking of mother earth. The only way, by which we can minimize the effects of the earthquake, is to design and build structures earthquake resistant.

This book is written to serve as a text book for a course on Earthquake Resistant Building Construction, for civil engineering students. The book covers the syllabus of the Board of Technical Education, Punjab, Harayana, Himachal Pradesh, J & K and other states also. The subject matter is explained in a simple and lucid language, with illustrations, to make the students understand the basics of earthquake resistant construction.

The book is divided into eight chapters, each dealing with a specific aspect of earthquake engineering. The first two chapters are devoted to understand the basics of earthquake and structural dynamics. Chapter 3 and 4 are focusing on the traditionally built constructions a their failure during earthquake and the features which improve their earthquake resistance. Chapter 5, 6 and 7 explain the behaviour of Reinforced Concrete Construction during earthquakes, their ductile detailing and estimation of earthquake forces. Chapter 8 discusses disaster management, especially for earthquakes. Various Indian Standard Seismic Codes are presented in the book and explained clause-wise for making the students aware of them.

In preparing the material of this book, I have drawn freely on the published literature on the subject, which is gratefully acknowledged. I am also thankful to my family, especially my husband Er. S.K. Rampal for his guided support, in bringing out this book.

Although lot of care has been taken, but it is natural that some errors might have crept in while making the book. I will appreciate if such errors are brought to my notice. Suggestions for improvement of the book are also welcome.

Neelam Sharma
AUTHOR

Contents

1

CHAPTER

Engineering Seismology

INSIDE THIS CHAPTER

1.1. INTRODUCTION

The word Earthquake is self explaining—the Earth-quakes that means the earth shakes and we feel the vibrations caused by these motions. Earthquakes are caused due to many reasons (later explained) but most commonly the term 'Earthquake' is used when shaking of the earth's surface is caused due to some disturbances occurring inside the earth.

Whenever the earth is disturbed, vibrations are produced. These vibration set out in all directions from the place of their origin. Wherever these vibrations travel, an earthquake is said to have taken place. These vibrations are more intense near their source. As the distance increases, these become feeble and slowly die out. More than 10,000 earthquakes occur every year. But most of them are not of great concern for civil engineers, only a few of them, having high intensity, are a cause of major concern. Some earthquakes can be very destructive and result in collapse of the structures, thus, resulting in heavy loss of life and damage to the buildings. Thus, it is necessary to study the earthquakes in detail and take every precautionary measure and protection, to minimize the loss of life and property. Since earthquakes are unpreventable and unpredictable, so we should design the structures earthquake resistant which means that they should withstand earthquake forces without much damage so that loss of life and property is minimised. The present chapter focusses on the various elements of earthquake – their causes, classification, seismic waves and their recording. The chapter, also explains the measuring of earthquake–size in terms of magnitude and intensity. At the end of the chapter seismic zoning is discussed and seismic map of India (1970 and 2016), are also given for reference.

1.2. SEISMOLOGY AND ITS IMPORTANCE

Earthquake is defined *as the shaking of earth's surface due to any reason which, results in release of large amount of energy.* The energy released during an earthquake is enormous. For example, the energy released during Bhuj earthquake (India 200) was about 400 times more than the energy released by the 1945 atom bomb dropped in Hiroshima. The word seism means earthquakes. Hence, *seismology or Earthquake Engineering is defined as the branch of science which deals with the study of earthquakes.* It includes causes and types of earthquakes, study of origin, propagation, recording and analyzing seismic waves.

Earthquakes are a major cause of concern for civil engineers because the effect of this disaster is measured in terms of extent of damage. This damage includes loss of lives and property both. Some body has rightly said that "Earthquakes do not kill people but it is the structures built by them that do so." Thus it is very necessary to make the structures earthquake resistant. So, the study of earthquake engineering is important for civil engineers to equip them with the basic knowledge of earthquakes, its effect on structures and various prnciples and techniques to be followed while designing and constructing earthquake resistant structures. In brief, the study of seismology or earthquake engineering is important because :

(*i*) It helps us in understanding the earthquakes, their nature and effect on our lives.

(*ii*) It helps us in designing and building earthquake resistant structures to minimize the loss of lives and property.

(*iii*) It also helps us in planning effective disaster management programme during earthquake by preparaing ourselves in a better way to reduce the effects of earthquake.

1.3. STRUCTURE OF EARTH

The earthquakes originates inside the earth, therefore it is essential to study the structure of the interior of the earth. The interior of the earth is divided into the following zones (Fig. 1.1)

1. Crust
2. Mantle
3. Core

1. Crust : The uppermost part of the earth is called as crust. It is solid and extends upto 60 km inside the earth. It mainly consists of rocks like granite, basalt etc. The density of crust varies from 2.5 to 3.0.

2. Mantle : Below the crust is mantle which is divided into two parts— outer and inner. The outer mantle extend upto 660 km and inner mantle upto 2900 km. The density of mantle varies from 3.4 to 5.5. This part is in semi-solid state and the approximate temperature inside the mantle is 3000°C.

3. Core : The innermost part of the earth is called as core. It is also divided into two parts—Outer and Inner. It is mainly liquid zone. In composition, it has Nickel and Iron. The outer core extend upto 5200 km and inner core from 5200–6370 km. The density of core varies from 11.0–16.0 and the temperature at the inner core is 6000°C. The boundary between crust and Mantle is called as MOHO.

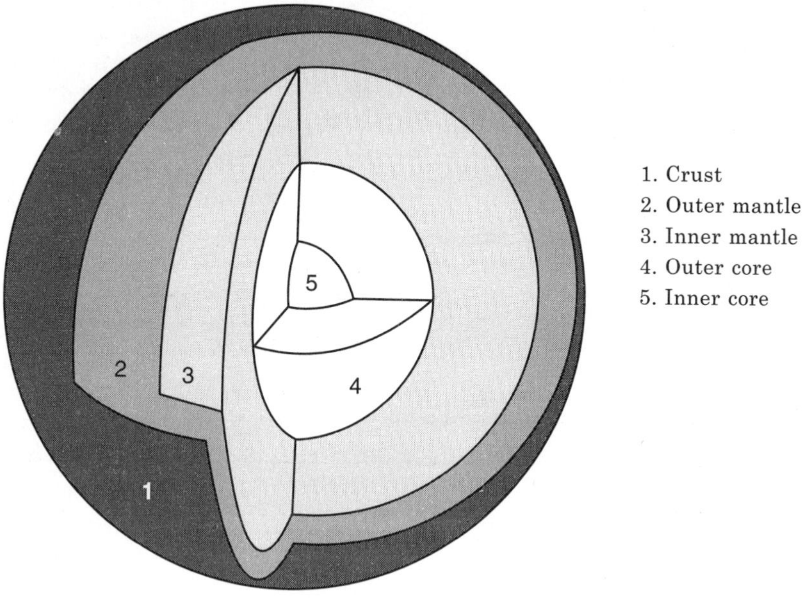

Fig. 1.1. Structure of the Earth.

1.4. PLATE TECTONICS

Theory of plate tectonics is based on the theory of continental drift which explains the origin of continents and oceans.

According to this theory, a long time ago, large amount of material masses joined together to form the earth. Large amount of heat was generated during this fusion. As the earth cooled down, the heavier material sank to the center and the lighter ones towards the top. The upper part formed is called as crust and the inner part is called as Core (ref : structure of the earth).

The high temperature and pressure difference between the crust and the core results in the convection currents (like flow of water molecules when heated up). These currents result in circulation of the earth's mass from crust to core and vice versa. The hot molten lava comes out and cold rock mass goes into the earth. This flow of material causes the Crust and some part of the Mantle to slide on the outer Core. This sliding of earth's mass takes place in different parts of the earth, called as *Tectonic plates*. The surface of earth consists of seven major tectonic plates (Fig. 1.2) and many small plates. These plates move in different directions and at different speeds. Thus, resulting in rising and sinking of the continents (formation of mountains and valleys) then Fig. 1.3.

In India, the raised beaches along the eastern and western coasts, and the submerged forests of Bombay are typical examples of plate movements. Some movements of these plates also occur tangential to the earth's surface and are very slow. These movements result in the formation of mountains (like Alps and Himalayas)

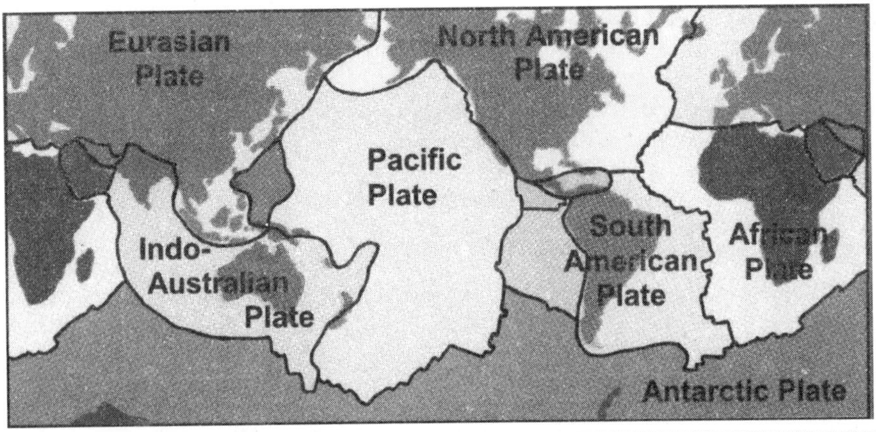

Fig. 1.2. Major Tectonic Plates of the earth.

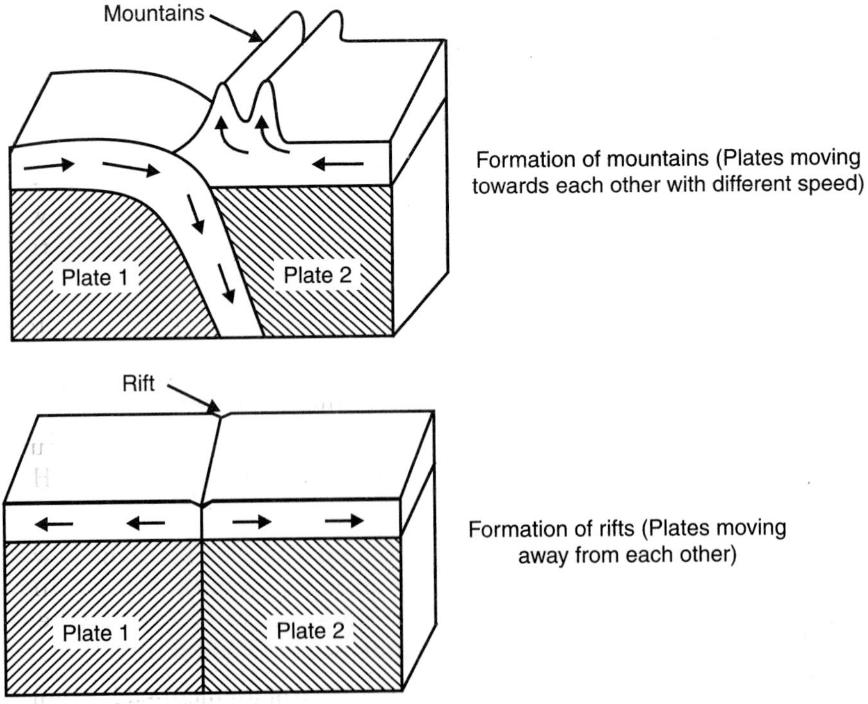

Fig. 1.3. Plate movements.

1.5. FAULTS

A fault is defined as a fracture or a crack, in the rocks, along which some definite movement has taken place. Movement in a fault can be of varying nature; it may be slow, sudden or repeated. The amount of movement may range from a few centimeters to several kilometers. Earthquakes are mainly caused due to sudden slip at the faults which results in shaking of the earth. This sudden slip also results in release of large amount of energy.

1.5.1. Basic Terminology of Faults (Refer Fig. 1.4)

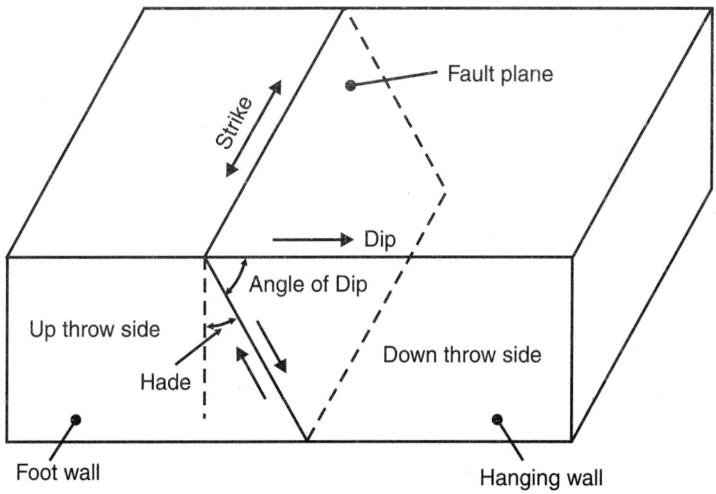

Fig. 1.4.

1. Fault Plane : The plane splitting the rock into two blocks along which movement occurs is known as fault plane. It may be horizontal or vertical.

2. Dip and strike of the fault : The inclination of the fault plane with the horizontal is called as dip of the fault. The direction perpendicular to dip is called strike of the fault.

3. Hade : It is the angle which a fault plane makes with vertical.

4. Throw and Heave : The total vertical displacement in a fault is known as throw of the fault and total horizontal displacement is known as Heave of the fault.

5. Slip of the Fault : The relative displacement of two points which were initially against each other is known as slip of a fault. The maximum displacement is called as net slip and along the strike is called as strike slip; displacement along the dip of the fault plane is called as dip slip.

6. Hanging Wall and Foot Wall : In the case of an inclined fault plane, the block which rests over the other is called as hanging wall and the underlying block is called as foot wall.

1.5.2. Types of Faults

Faulting is the chief cause of earthquakes. The sudden slip of fault produces vibrations in the earth's crust causing earthquakes. Thus, study of different types of faults on the basis of slip becomes necessary. On the basis of slip of fault, there are three types of faults.

(*a*) *Dip Slip Faults :* Those in which net slip is in the direction of the dip of the fault. (Fig. 1.5)

(*b*) *Strike Slip Fault :* Those in which net slip is in the direction of strike of the fault (Fig. 1.5).

(*c*) *Oblique Slip Faults :* Those in which net slip is neither along dip nor along strike of the fault; rather it is oblique.

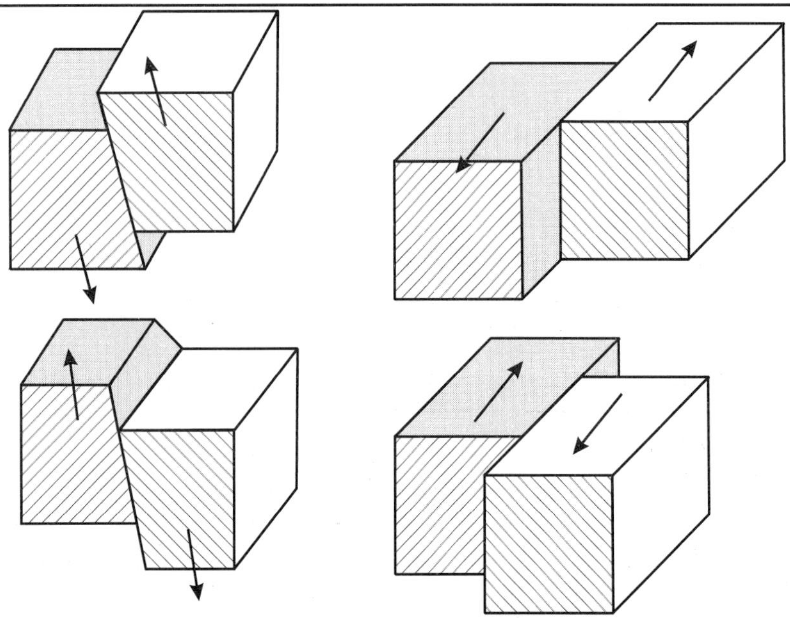

(*a*) Dip Slip Faults (*b*) Strike Slip Faults

Fig. 1.5. Type of faults.

1.6. CAUSES OF EARTHQUAKES

Earthquakes are primarily caused due to two reasons :

1. Natural disturbances
   (*a*) Volcanic causes
   (*b*) Tectonic causes
2. Artificial disturbances.

1. Natural Disturbances : The natural disturbances which causes earthquakes are following :

(*a*) *Volcanic Causes :* Volcanic activity keeps on taking place in several parts of the world. Very often, it produces sudden out burst or explosions.

This impact is sometimes strong enough to produce vibrations in the nearby areas. People, living in Japan and Italy, have experience this type of earthquake frequently. These earthquakes are not very deep and of mild intensity. The damage caused due to this type of earthquake is confined within a few kilometers. All volcanic eruptions don't produce earthquake.

(*b*) *Tectonic Causes :* Tectonic causes are those which occur inside the earth. According to the theory of plate tectonics, the Crust (Outermost layer of Earth) is made up of rocks and is divided into many plates. These plates are constantly in motion in different directions and with different speeds. These rocks or plates have strain energy stored in them.

The sudden slip at the fault (A sudden movement along a fracture or crack in the rocks) results in the release of strain energy in the form of seismic waves which causes the shaking of the earth. *The earthquake which occur due to sudden release of strain energy as a result of sudden slip of fault is called as tectonic earthquake.* This slip may occur along the preexisting crack/ fault or along a new fault plane.The mangitude of these earthquake depends upon the amount of energy released. More than 90% of total earthquakes are because of tectonic reasons. Bhuj (2001) and Latur (1993) were tectonic earthquake.

2. Artificial Disturbances : Sometimes the surface of the earth vibrates due to man made or artificial disturbance. These vibrations are very mild and affect the surrounding area only. The earthquakes of mild intensity are caused due to these external man made agencies. Some of the artificial disturbances causing earthuakes are listed below :

(*i*) Nuclear tests and explosions.

(*ii*) Mining blasts in the mining area.

(*iii*) A massive landslide along hill slopes caused because of deforestation.

(*iv*) Large and deep excavations.

(*v*) Vibrations induced due to heavy machinery used in industries or movement of heavy vehicles.

The vibrations or shaking caused due to above reasons is very minor and limited to small areas only. All these causes occur over the earth's surface so these are also called as surface causes.

1.7. TERMINOLOGY

Some of the basic terminology related with earthquake is given below [refer Fig. 1.6]

1. Hypocenter : The point inside the surface of the earth from where the earthquake originates is termed as hypocentre. This is also known as focus. Hypocenter is the point on the fault where the slip or the movement starts.

2. Epicenter : The point on the earth's surface, vertically above the hypocenter is called as epicenter.

3. Focal depth : The depth of focus from the epicenter is called as focal depth. It is an important parameter in determining the extent of damage of an earthquake. The focal depth of most of the major earthquakes causing lot of damage is shallow.

4. Epicentral distance : Distance from epicenter to point of any interest is called as epicentral distance. As the epicentral distance is increased the effect of earthquake becomes less. The damage is maximum at the epicenter and surrounding area.

5. Fore shocks and after shocks : A fault rupture is not a simple one time movement that produces "the earthquake". In fact we never have just one earthquake. The fault movement generates a series of earthquakes. Out of these the biggest one is called as the earthquake (Main shock). The smaller size earthquakes that takes place before a big earthquake are called foreshocks and those after the big earthquake are called after shocks.

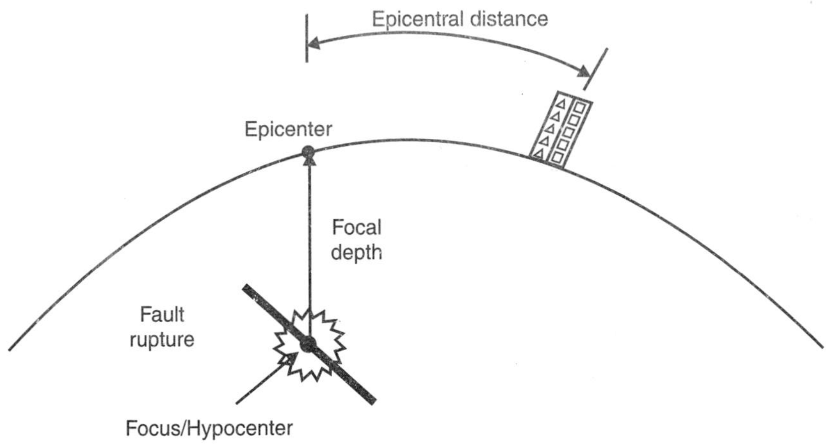

Fig. 1.6. Basic terminology of earthquake.

1.8. SEISMIC WAVES

When the earth shakes, it relases energy in the form of seismic waves which travel inside the earth and also on the surface of the earth. These waves obey the laws of reflection and refraction. Seismic waves are of the following two types :

(*a*) Body Waves
(*b*) Surface Waves

(*a*) Body waves : The seismic waves which originate inside the earth and travel through the various layers of the earth in all directions are called as body waves. Body waves are of following two types :

(*i*) Primary Waves
(*ii*) Secondary Waves

(*i*) **Primary Waves :** Primary waves or P-waves moves in a push and pull fashion like sound waves. Thus, the material particles undergoes compression (push) and extension (pull) as shown in Fig. 1.7. P-waves radiate outward from the source like a gas filled balloon. They are also termed as longitudinal waves.

Properties of P Waves

(*i*) P waves are the fastest seimic waves and thus first to reach at the recording station.

(*ii*) P waves can travel through any material be it solid, liquid or gas.

(*iii*) Their speed depends upon the density and compressibility of the material through which they are passing.

(*iv*) These are longitudinal waves like sound waves which move in push and pull fashion.

(*v*) P waves make loud thunderous noise near the epicentre.

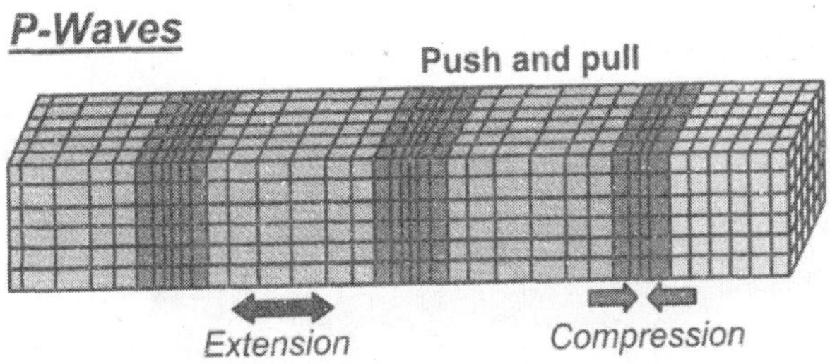

Fig. 1.7. Primary Body Waves.

(*ii*) **Secondary waves :** Secondary waves or S-waves are those body waves which propagate by vibrating the material particles at right angle to the direction of propagation. These waves are transverse in nature like light waves. There are also called as shear waves. (Fig. 1.8)

Properties of Secondary Waves

(*i*) S-waves are slower than P-waves and hence second to reach at the recording station.

(*ii*) S-waves can travel through solids only because liquids and gases do not have shear strength.

(*iii*) S waves are transverse in nature like light waves.

(*iv*) The velocity of S waves depends upon the density and shear strength of the material through which they pass.

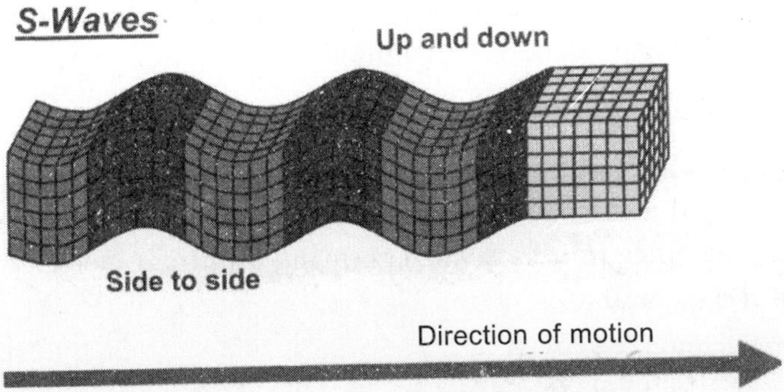

Fig. 1.8. Secondary Body Waves.

(b) Surface waves : The seismic waves which travel along the surface are called as surface waves. These waves do not propagate deep inside of the surface of the earth. These are also called as long waves. These waves can be visualized as the ripples caused by throwing a rock into water. Surface waves are the slowest, thus last to reach the recording station. Surface waves shake the ground surface and are responsible for all the damage occurring on the earth's surface. These are most destructive in nature. Surface waves are of two types :

(i) Love waves
(ii) Rayleigh waves.

(i) **Love waves :** These waves were detected by seismologist A.E.H. Love in 1911 and thus are named as Love waves.

Properties of Love Waves

(i) Love waves cause surface motion similar to S-waves, but with no vertical component, thus the motion is from side to side in a horizontal plane roughly parallel to the earth's surface. (Fig. 1.9)

(ii) Love waves travel faster than Rayleigh waves and like S-waves they do not move through liquid or air.

(iii) Love waves along with S-waves cause maximum damage to the structures.

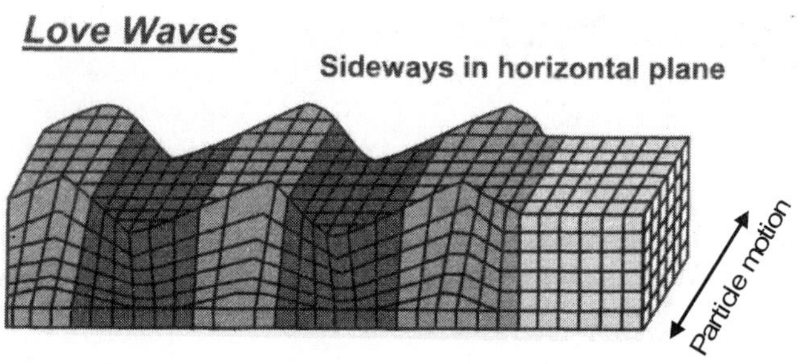

Fig. 1.9. Surface waves (Love waves).

(ii) **Rayleigh waves :** These waves are first discovered by Rayleigh in 1885. In these waves the material particles vibrate in a backward elliptical path in the vertical plane.

Properties of Rayleigh Waves

(i) The shaking produced by Rayleigh waves causes both vertical and horizontal movement.

(ii) They advance in a backward rotating elliptical motion as shown in Fig. 1.10.

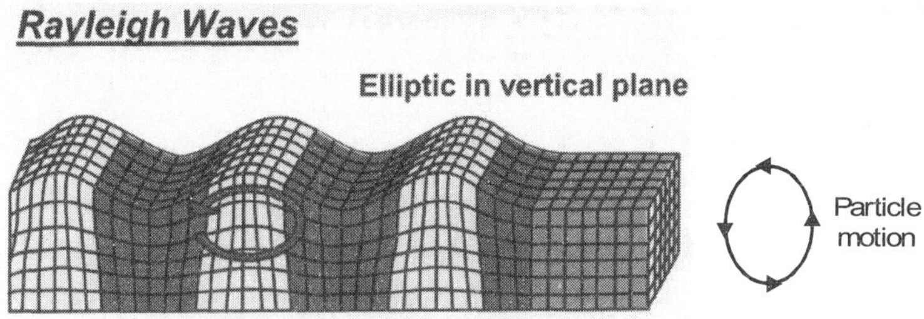

Fig. 1.10. Surface waves (Rayleigh waves).

The study of seismic waves and their records is very important in seismology. They not only help us in assessing the magnitude, intensity and location of the earthquake but also give us information about the interior of the earth. For example, the S waves help us in knowing that outer core is liquid because as they pass through the mantle, they disappear. Similarly, movement of P-waves into the core (outer to inner indicates an increase in velocity), suggest that the inner core is solid.

1.9. SEISMOGRAPH

Shaking of ground on the Earth's surface is caued by seismic waves. The instrument which measure earthquake shaking and record them is called as Seismograph. A typical seismograph has following three parts :

(*a*) Sensor
(*b*) Recorder
(*c*) Timer

(*a*) **Sensor :** The sensor is used to sense or detect vibrations. It consists of a pendulum mass, string, magnet and support.

(*b*) **Recorder :** The recorder is used to record the seismic vibrations on a paper. It consists of a drum, pen and chart paper. The chart paper is wrapped on the drum.

(*c*) **Timer :** The timer is the motor that rotates the drum having chart paper at constant speed.

Working of a Seismograph

The construction of a typical seismograph is shown in Fig. 1.11. The simple pendulum (a string and mass) is suspended freely but a magnet is tied around the string to dampen the oscillation (i.e. to control or reduce the amplitude of oscillations). A pen is attached at the tip of the pendulum which marks the movement or oscillations of the pendulum on the chart paper wrapped around the rotating drum. Whenever there is earthuake, the vibrations are produced and the pendulum vibrates simultaneously thus recording these vibrations on the paper with the pen. Since the earthquake motion is two dimensional, one such instrument is required in each perpendicular direction to record both the components of motion.

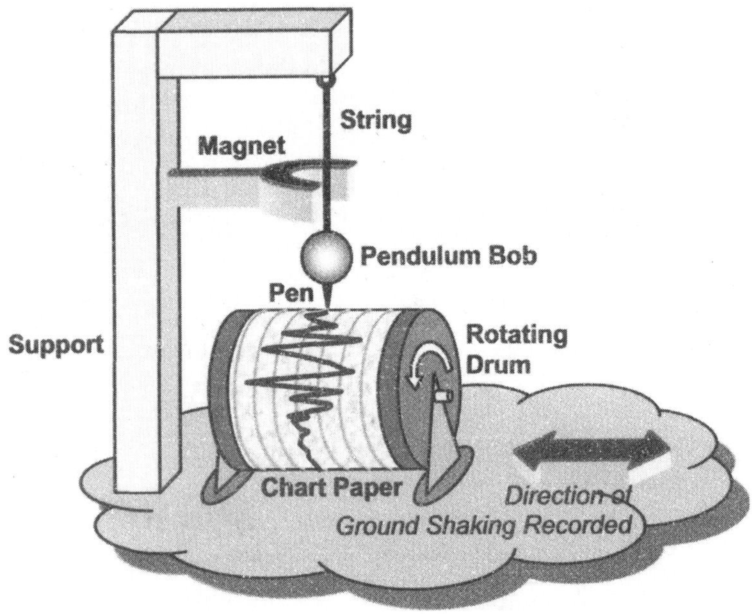

Fig. 1.11. Seismograph.

The seismographs can record only mild and weak vibrations which are not of high intensity. The strong motion or high intensity vibrations are recorded by special instruments called as **accelerographs.**

Some instruments do not have a timer device (i.e. the drum holding the chart paper doesn't rotates). Such instruments provide only the maximum extent of motion during the earthquake. They are called **Seismoscopes.**

Nowadays digital instruments are used which records the ground motion on the memory of inbuilt computer.

1.10. SEISMOGRAM

The record of an earthquake is called as seismogram. In a seismogram, the earthquake waves are recorded in the order, they arrive at the recording station. P-waves being the fastest, recorded first, followed by S-waves and then slowest L-waves (Fig. 1.12). Using the time, the various waves take to reach the station, the location of the epicenter and hypocenter, focal depth etc. can be determined.

As already explained, all earthquakes are not of concern for Civil engineers. Only earthquakes causing strong motion are relevant to Civil engineers. For this, strong motion recording instruments called accelerographs are used. The motion of the ground can be described in terms of displacement, velocity and acceleration. *The variation of ground acceleration with time, recorded at a point on ground, during earthquake is called an* **accelerogram.** An accelerogram is shown in Fig. 1.13(*a*). Figure 1.13(*b*) shows an accelerogram showing three components of the earthquake motion.

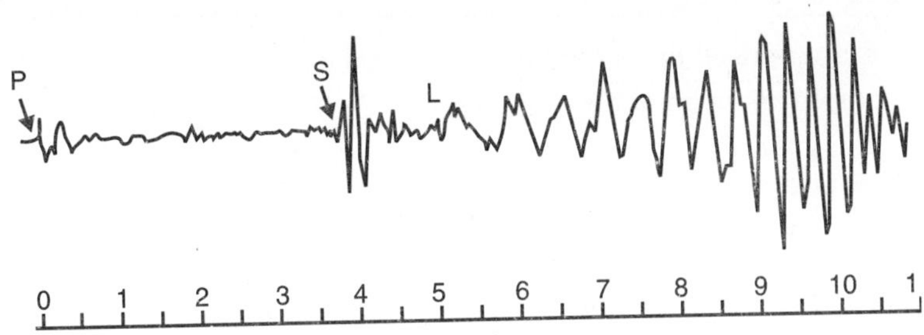

Fig. 1.12. Seismogram.

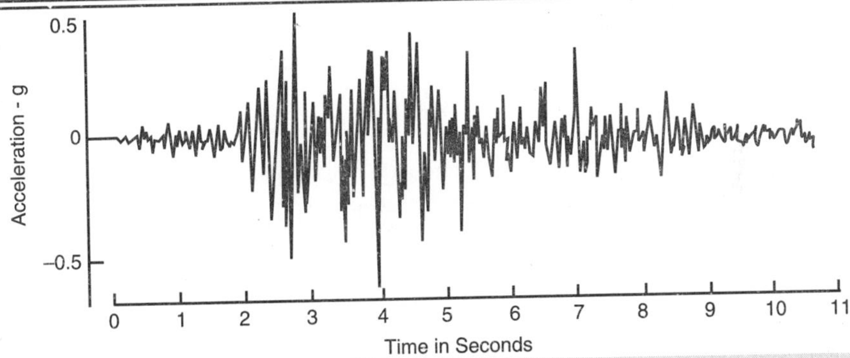

Fig. 1.13(a). Accelerogram.

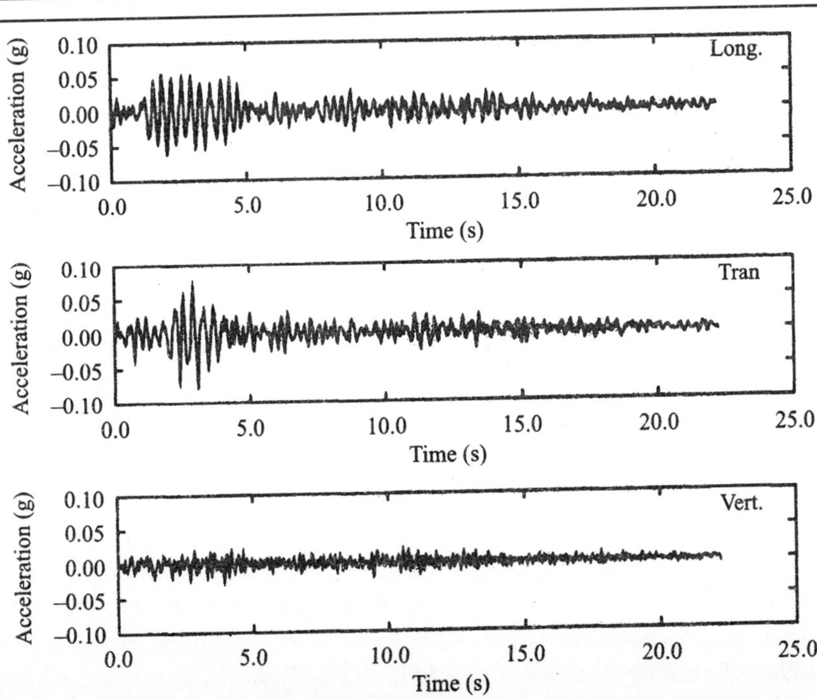

Fig. 1.13(b). Three components of the earthquake motion.

1.10.1. Use of Seismogram

Seismogram is used for the following purposes :

1. Seismogram is used for determining the epicenter of the earthquake.
2. Seismograms are the records of the earthquakes, thus they are used for obtaining the seismic parameters which are used in design of structures and also for identifying the seismic zones.
3. Seismogram also helps us in studying the seismic waves their nature and propagation which helps in assessing the time and severity of the earthquake.

1.11. EARTHQUAKE SIZE

Earthquake size is defined in terms of two things :

1. Magnitude
2. Intensity.

Earthquake size is a measure of the qualitative and quantitative effects of the vibrations produced by the earthquake.

1.11.1. Magnitude

Magnitude is a quantitative estimate of the earthquakes size. It is a measure of amount of energy released during the earthquake. The magnitude of an earthquake is generally measured on **Richter Scale.**

Dr. Charles Richter noticed that the Seismic waves provide a good estimation of their magnitude. He studied seismic waves and found that

(*a*) At the same distance, seismograms of larger earthquakes have bigger wave amplitude than those of smaller earthquakes.

(*b*) For a given earthquake, Seismograms at farther distances have smaller wave amplitude than those at closer distance. It means that wave amplitude decreases with increase in epicentral distance.

His findings led him to propose the new commonly used **Richter Scale.** According to Ritcher, the *Magnitude of an earthquake can be defined as Logarithm (base 10) of the maximum amplitude A (in microns = 1/1000 mm) of the ground motion on a standard Seismograph at a distance of 100 km from the epicentre.* This scale is also called as Local Magnitude Scale. In this scale, whole numbers are used to describe magnitude. For every 1 unit increase in magnitude the wave amplitude is increased 10 times and about 31 times energy is released. For example energy released in a M6 earthquake is about 31 times that released in M5 earthquake and about 900 times that released in a M4 earthquake. Most of the energy released goes in the form of heat and fracturing the rocks. Only a little amount goes in the form of Seismic waves that causes shaking of the ground. A relationship between energy released by an earthquake and its magnitude is given by Richter as follows :

$$\log_{10} E = 11.4 + 1.5\,M$$

where E is the energy released and M is the magnitude on Richter's Scale.

There are other magnitude scales also like the Body Wave Magnitude, Surface Wave Magnitude and Wave Energy magnitude. These are all numerical scales and have no upper and lower limits i.e. the magnitude of a small earthquake can be zero or even negative. Earthquakes are often classified based on the magnitude. (Table 1.1)

TABLE 1.1. Earthquake classification based on their magnitude and their global occurrence.

Group	Magnitude	Annual average number
Great	8 and higher	1
Major	7–7.9	18
Strong	6–6.9	250
Moderate	5–5.9	800
Light	4–4.9	~6,200 (approx)
Minor	3–3.9	~49,000 (approx.)
Very minor	< 3.0	~9,00,000 per day

1.11.2. Intensity

Intensity is a qualitative measure of the earthquake. It is a way of measuring or rating the effects of an earthquake at different sites. It is generally expressed as Roman Numerals (I, II, III, ...). The assignment of Intensity does not require any instrumental records. It depends very much upon the visual observations and experiences of the people. Intensity data are very much useful for the development of Seismic risk map of an area or country. These maps have got various uses like planning safe sites for important structures like large dams and nuclear power plants and determination of historic seismicity of an area etc. The intensity of an earthquake depends on the following factors :

1. **Earthquake Magnitude :** The relation between magnitude and intensity is very clear—the bigger the earthquake (more magnitude), the higher is the damage to lives and property.

2. **Distance from Hypocenter or Epicenter :** The relation between epicentral distance and damage is straight forward. It is observed that the places near the epicenter suffer more damage as compare to those which are farther.

3. **Type of Foundation Material :** The type of rock or soil are of importance while evaluating earthquake effects because it can help us in assessing the extent of damage possible. For example, hard rock foundations can vibrate more by P & S-waves from a near by epicenter and soft soils can vibrate more by surface waves from a distant earthquake and often result into landslides.

4. **Building Style :** The loss of lives is not because of shaking of the earth but because of collapse of buildings, bridges and other structures. The building style includes the design of building, building materials, height and shape of building etc. For example, P & S waves can cause higher damage to brick or stone buildings especially the shorter ones. The surface waves can cause great damage to tall buildings built on soft soil.

5. **Duration of Shaking :** The duration of shaking is a significant factor for assessing damage suffered and lives lost. If the shaking duration is more obviously the damage is more.

Based on the above factors two scales are commonly used—Modified Mercalli Intensity (MMI) Scale and the MSK Scale. Both scales are quite similar and range from I (least severity) to XII (most severe). The Intensity Scales are developed keeping in view the following things :

(*a*) the experiences of people

(*b*) performance of buildings

(*c*) changes in natural surroundings.

1.11.2.1. MSK Intensity Scale

MSK Intensity Scale takes into account the type of structure (Table 1.2), percentage of damage to each type of structure (Table 1.3), grade of damage to different type of structures (Table 1.4) and details of intensity scale (Table 1.5).

TABLE 1.2. Type of Structures (buildings)

Type of structures	Definitions
A	Buildings in fieldstone, rural structures, unburnt-brick houses, clay houses.
B	Ordinary brick buildings, buildings of the concrete block and prefabricated type, half-timbered structures, buildings in natural hewn stone.
C	Reinforced buildings, well-built wooden structures.

TABLE 1.3. Definition of Quantity

Quantity	Percentage
Single, few	About 5
Many	About 50
Most	About 75

TABLE 1.4. Classification of damage to buildings

Grade	Definitions	Descriptions
Gl	Slight damage	Fine cracks in the plaster; fall of small pieces of plaster.
G2	Moderate damage	Small cracks in walls; fall of fairly damaged large pieces of plaster, pantiles slip off; cracks in chimneys; parts of chimney brakes.
G3	Heavy damage	Large and deep cracks in walls; fall of chimneys.
G4	Destruction	Gaps in walls; parts of buildings may collapse; separate parts of the building lose their cohesion; and inner walls collapse.
G5	Total damage	Total collapse of buildings.

TABLE 1.5. Details of MSK intensity scale.

Intensity		Descriptions
I	Not noticeable	The intensity of the vibration is below the limit of sensibility; the tremor is detected and recorded by seismographs only.
II	Scarcely noticeable (very slight)	Vibration is felt only by individual people at rest in houses, especially on upper floors of the buildings.
III	Weak, partially observed only	The earthquake is felt indoors by a few people, outdoors only in favourable circumstances. The vibration is like that due to the passing of a light truck. Attentive observers notice a slight swinging of hanging objects, somewhat more heavily on upper floors.
IV	Largely observed	The earthquake is felt indoors by many people, outdoors by few. Here and there people awake, but no one is frightened. The vibration is like that due to the passing of a heavily loaded truck. Windows, doors and dishes rattle. Floors and walls crack. Furniture begins to shake. Hanging objects swing slightly. Liquids in open vessels are slightly disturbed. In standing motorcars the shock is noticeable.
V	Awakening	(*a*) The earthquake is felt indoors by all, outdoors by many. Many sleeping people awake. A few run outdoors. Animals become uneasy. Buildings tremble throughout. Hanging objects swing considerably. Pictures knock against walls or swing out of place. Occasionally pendulum clocks stop, Unstable objects may be overturned or shifted. Open doors and windows are thrust open and slam-back again. Liquids spill in small amounts from well-filled open containers. The sensation of vibration is like that due to heavy object falling inside the buildings. (*b*) Slight damages in buildings of Type A are possible. (*c*) Sometimes changes in flow of springs.
VI	Frightening	(*a*) Felt by most indoor and outdoor people. Many people in buildings are frightened and run outdoors. A few persons lose their balance. Domestic animals run out of their stalls. In few instances dishes and glassware may break, books fall down. Heavy furniture may possibly move and small steeple bells may ring.

continued on the next page

		(*b*) Damage of Grade I is sustained in single buildings of Type B and in many of Type A. Damage in a few buildings of Type A is of Grade 2. (*c*) In few cases cracks up to widths of 1 cm is possible in wet ground; in mountains occasional landslips; change in flow of springs and in level of well water are observed.
VII	Damage of buildings	(*a*) Most people are frightened and run outdoors. Many find it difficult to stand. The vibration is noticed by persons driving motorcars. Large bells ring. (*b*) In many buildings of Type C damage of Grade I is caused; in many buildings of Type B damage is of Grade 2. Most buildings of Type A suffer damage of Grade 3, a few of Grade 4. (*c*) In single instances landslips of roadway on steep slopes; cracks in roads; seams of pipelines damaged; cracks in stone walls.
VIII	Destruction of buildings	(*a*) Fright and panic; also persons driving motorcars are disturbed. Here and there branches of trees break off. Even heavy furniture moves and partly overturns. Hanging lamps are damaged in part. (*b*) Most buildings of Type C suffer damage of Grade 2, and a few of Grade 3. Most buildings of Type B suffer damage of Grade 3, and most buildings of Type A suffer damage of Grade 4. Many buildings of Type C suffer damage of Grade 4. Occasional breaking of pipe seams. Memorials and monuments move and twist. Tombstones overturn. Stone-walls collapse. (*c*) Small landslips in hollows and on banked roads on steep slopes; cracks in ground up to widths of several cm. Water in lakes becomes turbid. New reservoirs come into existence. Dry wells refill and existing wells become dry. In many cases change in flow and level of water is observed.
IX	General damage to buildings	(*a*) General panic; considerable damage to furniture. Animals run to and fro in confusion and cry. (*b*) Many buildings of Type C suffer damage of Grade 3, and a few of Grade 4. Many buildings of Type B show damage of Grade 4, and a few of Grade 5. Many buildings of Type A suffer damage of Grade 5. Monuments and columns fall. Considerable damage to reservoirs; underground pipes partly broken. In individual cases railway lines are bent and roadway damaged.

continued on the next page

		(*c*) On flat land overflow of water, sand and mud is often observed. Ground cracks to widths of up to 10 cm, on slopes and river banks more than 10 cm; furthermore a large number of slight cracks in ground; falls of rock, many landslides and earth flows; large waves in water. Dry wells renew their flow and existing wells dry up.
X	General destruction of buildings	(*a*) Many buildings of Type C suffer damage of Grade 4, and a few of Grade 5. Many buildings of Type B show damage of Grade 5; most of Type A have destruction of Grade 5; critical damage to dams and dykes and severe damage to bridges. Railway lines are bent slightly. Underground pipes are broken or bent. Road paving and asphalt show waves. (*b*) In ground, cracks up to widths of several cm, sometimes up to 1 m. Parallel to water course occur broad fissures. Loose ground slides from steep slopes. From river-bank and steep coasts, considerable landslides: are possible. In coastal areas, displacement of sand and mud; change of water level in wells; water from canals, lakes, rivers, etc., thrown on land. New lakes; occur.
XI	Destruction	(*a*) Severe damage even to well built buildings, bridges, water dams and railway lines; highways become 1 useless; underground pipes destroyed. (*b*) Ground considerably distorted by broad cracks and fissures, as well as by movement in horizontal and vertical directions; numerous landslips and falls of rock. The intensity of the earthquake requires to be investigated specially.
XII	Landscape changes	(*a*) Practically all structures above and below the ground are greatly damaged or destroyed. (*b*) The surface of the ground is radically changed. Considerable ground cracks with extensive vertical and horizontal movements are observed. Falls of rock and slumping of river-banks over wide areas, lakes are dammed; waterfalls appear, and rivers are deflected. The intensity of the earthquake requires to be investigated specially.

1.11.2.2. MMI Scale

The details of MMI Scale (Modified Mercalli Scale) are given below :

TABLE 1.6. Details of Modified Mercalli Scale (MMI).

Intensity	Description
I	People do not feel any earth movement.
II	Felt by a few people only especially those who are at rest on upper floors.
III	Shaking felt indoors, hanging objects swing. Vibrations are like those from the passing of heavy truck. Standing automobiles may rock slightly. People outdoors might not feel that an earthquake is occurring.
IV	Doors, windows and dishes rattle, standing cars rock, trees shake. Most people indoor feels movement, hanging objects swing. The earthquake feels like a heavy truck hitting the walls. A few people outdoors may feel movement light sleepers may be awakened.
V	Felt indoors by nearly everyone, outdoors by many sleepers awake, doors swing, dishes, glassware and windows are broken. Minor cracking of plaster, liquid spills, tall objects sways. Trees might shake.
VI	Felt by all many frightened and run outdoors dishes glassware and windows break in considerable quantities, objects fall from shelves, picture fall off walls, furniture moves, Trees and bushes shake, people have trouble walking, plaster in walls might crack. No structural damage but slight damage in poorly constructed buildings.
VII	Frightens all, difficult to stand, large bells ring, plaster, bricks and tiles fall, people have difficulties in standing, drivers feel their cars shaking, some furnitures break, damage is slight to moderate in well-built buildings, considerable in poorly built buildings.
VIII	Car steering affected, chimneys fall, branches break, cracks in wet ground, drivers have trouble steering. Houses that are not bolted down the foundation, might shift. Tall structures such as towers and chimneys might twist and fall. Well built buildings suffer slight damage. Poorly built structures suffer severe damage, Tree branches break, Hill sides might crack if the ground is wet. Water levels in wells might change.
IX	General panic, damage to foundations, sand and mud bubble from ground. Well built buildings suffer considerable damage. Houses may move off their foundations, some underground pipes are broken. The ground cracks. Reservoirs suffer serious damage.
X	Most buildings destroyed, large landslides, water thrown out off the rivers. Most buildings and their foundations are destroyed. Some bridges are destroyed. Dams are seriously damaged. Large landslide occur. The ground cracks in large areas. Railroad tracks are bent slightly.
XI	Railway tracks are badly bent, roads break up, large cracks appear in ground, rocks fall. Most buildings collapse. Some bridges are destroyed. Underground pipes are broken.
XII	Total destruction, waves seen on ground surface, river course altered, vision distorted. Almost everything is destroyed object are thrown into the air. The ground moves in waves or ripples. Large amounts of rock may move.

1.11.3. Magnitude Versus Intensity

- Magnitude is a quantitative estimate of earthquake while intensity is qualitative.
- Magnitude of an earthquake is a measure of its size in the form of amount of strain energy released by the fault rupture. Intensity is an indicator of the severity of the earthquake caused due to shaking of ground at a given location.
- Magnitude of an earthquake is a single value. On the other hand, intensity is different at different locations, thus, many values. Clearly, the severity of shaking is much higher near the epicenter than farther away.

TABLE 1.7. Comparison of Magnitude, Intensity and Acceleration.

Richter Magnitude	Mercalli Intensity	Acceleration (%g)
2 and less	I-II	Usually not felt by people < 0.1–0.19
3	III	Felt indoors by some people, 0.2–0.49
4	IV-V	Felt by most people, 0.5–1.9
5	VI-VII	Felt by all, building damage, 2–9.9
6	VII-VIII	People scared, moderate damage, 10–19.9
7	IX-X	Major damage, 20–99.9
8 and up	XI-XII	Damage nearly total, Over 100 = 1 g

1.11.4. Magnitude and Intensity in Seismic Design

An earthquake of a particular magnitude value causes different shaking intensities at different locations. The damage induced in buildings at these locations is different. Thus, it is a particular intensity level of shaking that buildings and structures are designed to resist. *The Peak Ground Acceleration (PGA) is the maximum ground acceleration experienced by the ground during shaking, is one of the way to express the severity of shaking.* Approximate empirical formulae are available to correlate MMI and the PGA that may be experienced (Table 1.7)

1.12. ISOSEISMALS

An imaginary contour line joining places of equal earthquake intensity is called as Isoseismal and a map showing different isoseismals for a particular earthquake is called as an isoseismal map (Fig. 1.14). The intensity is maximum near the earthquake epicenter and decreases with distance and at large distances earthquake is not felt.

1.12.1. Uses of Isoseismals

(*i*) Earthquake isoseismal maps provide important documents of seismic effects of earthquakes.

(*ii*) Isoseismal maps of past earthquakes also help us to understand the nature of the earthquakes in a particular region.

(*iii*) They are also used for long term planning of structures and for development of seismic zoning maps.

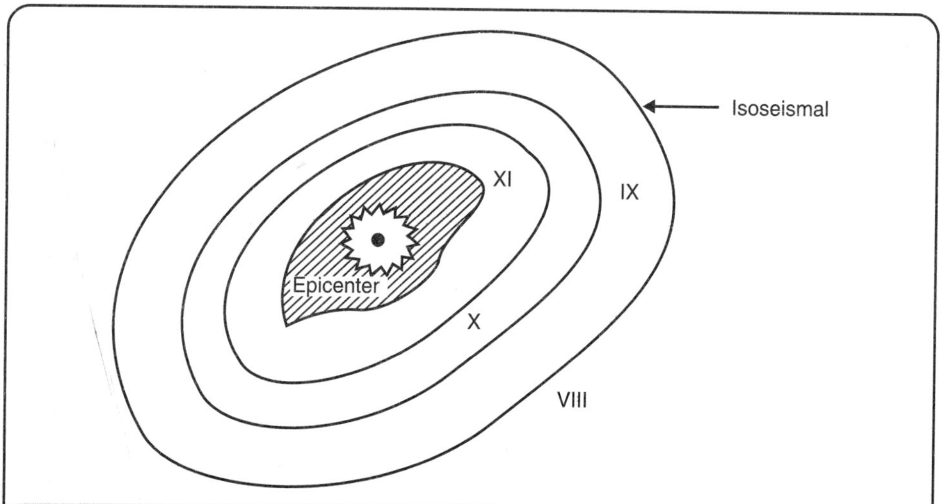

Fig. 1.14. Isoseismal map.

1.13. CLASSIFICATION OF EARTHQUAKES

Earthquakes are classified according to various criteriae explained below :

1. Based on location
2. Based on focal depth
3. Based on their cause
4. Based on their magnitude
5. Based on the epicentral distance

1. Based on Location : On the basis of location of occurrence of earthquake, they are classified as :

(*a*) Interplate earthquake

(*b*) Intraplate earthquake.

(*a*) *Interplate earthquake :* Most of earthquakes, in the world, occur along the boundaries of the tectonic plates and are called Interplate earthquakes. It is seen that epicenters of about 99% of earthquakes are in the narrow zones of interplate boundaries. For example, Assam earthquake (India, 1897), Gujarat 2001 and Indian Ocean 2012.

(*b*) *Intraplate earthquake :* The earthquakes occurring within a plate itself away from the plate boundaries are called as Intraplate earthquakes. For example, Latur (India, 1993), and Klahoma, 2016 was an intraplate earthquake.

2. Based on Focal Depth : The earthquakes are classified into following three types based on the focal depth :

(*a*) Shallow earthquake

(*b*) Intermediate earthquake

(*c*) Deep earthquake.

(*a*) *Shallow earthquake :* The earthquake whose focal depth (distance from focus to epicenter) is less than 70 km are called Shallow depth earthquakes. Nearly 80% of earthquakes are shallow, and are of great concern for civil engineers.

(*b*) *Intermediate earthquake :* Those earthquakes in which focal depth is from 70 to 300 km are called as Intermediate earthquake.

(*c*) *Deep earthquake :* The earthquake whose focal depth is greater than 300 km is called as deep earthquake.

3. Based on Their Cause : The earthquakes are classified into following two types based on their cause of origin :

(*a*) Non Tectonic earthquake

(*b*) Tectonic earthquake

(*a*) *Non-Tectonic earthquakes :* The earthquakes which are caused due to non-tectonic reasons like volcanic activities and man-made reasons are called as non-tectonic earthquakes. Non tectonic earthquakes are of following two types :

(*i*) *Volcanic earthquakes :* The earthquakes which are caused due volcanic activities are called as volcanic earthquakes.

(*ii*) *Artificial Earthquake :* These are caused by the artificial disturbance which are man made like nuclear power testing, blasts, mining, large dams, deforestation and land slides etc.

(*b*) *Tectonic earthquakes :* The earthquakes which are of tectonic origin and are caused due to sudden slip in the fault of the tectonic plates of the earth are called as tectonic earthquakes.

4. Based on the Magnitude of Earthquake : The earthquakes are classified as given below on the basis of their magnitude.

TABLE 1.8. Classification of earthquake based on Richter's magnitude.

Type	Magnitude
Great earthquake	8 and more
Major earthquake	7-7.9
Strong earthquake	6-6.9
Moderate earthquake	5-5.9
Light earthquake	4-4.9
Minor earthquake	3-3.9
Micro earthquake	< 3.0

5. Based on Epicentral Distance : On the basis of epicentral distance (affected area), the earthquakes are classified into following three types :

(*a*) Local earthquake
(*b*) Regional earthquake
(*c*) Teleseismic earthquake.

(*a*) *Local earthquake :* The affected area is very less within 1° of the epicenter of the earthquake.
(*b*) *Regional earthquake :* The affected area due to the earthquake is within 1° to 10° of the epicenter of earthquake.
(*c*) *Teleseismic earthquake :* The affected area is greater than 10° of the epicenter of the earthquake.

1.14. SEISMIC ZONING

The seismicity or earthquake activity at a place is assessed by, its distance from the faulted rocks, and from the past records. It varies in different parts of a country or region. It is seen from the past records that earthquake activity (their severity, occurrence etc.) are more or less same in a particular area or zone. Therefore, a country can be divided into zones of similar seismic activity. *Seismic zoning can be defined as dividing a country or region into smaller zones on the basis of their seismicity or earthquake activity (Intensity of damage).*

The seismic zoning of a country mainly depends upon the following :

(*a*) The seismic history of a region. It means the detail study of the earthquakes that have occurred in past, their characteristics like magnitude, intensity and extent of damage etc.
(*b*) The tectonic features of the place which could cause an earthquake in future.

1.14.1. Use of Zoning Map

The zoning maps are very useful for design of earthquake resistant structures. We can easily come to know about the various design forces depending upon the place (site's location) because the forces for which structures are designed depend upon the seismicity of that place. However, for important structures, it is necessary to take geological and seismological observations for the site to get design forces caused by the earthquakes. These should be studied alongwith historical data of earthquakes and then various seismic factors for design could be determined.

The zoning maps also provide information of expected earthquake effects in terms of Maximum Intensity of earthquake (MMI) or ground motion characteristics at a particular region.

1.15. TECTONIC FEATURES OF INDIA

India lies at the north-western end of the Indo-Australian plate. On this plate, India, Australia, a major part of the Indian Ocean and other smaller countries are located. This plate is colliding against other neighbouring plates. The three major tectonics plates (Fig. 1.15) of India are following :

1. Northern Himalayas
2. Plains of the Ganges and other rivers.
3. The peninsular region.

1. Northern Himalayas : The Himalayan ranges comprises of highly seismic zone which is very unstable. All along the foothills of the Himalayas, stretching from Kashmir to Assam is a long thrust fault known as the Great Boundary fault. This fault is highly prone to earthquakes and is marked as highly seismic zone. India's earthquake history shows that most of the destructive earthquakes ranking as severe earthquakes of the world, took place in this zone, which were interplate earthquakes. Table 1.9 shows various earthquakes that have occured in this region.

2. Plains of the Ganges and other rivers : Leaving aside the zones of Great Boundary Fault, other regions of Himalayan belt and Indo-Gangetic plains are moderately seismic areas. A study about the formation of Indo-Gangetic plains indicates that this zone is made of the alluvial deposits which are laid into a great depression present in front of the Himalayan ranges. This zone is still in an unstable condition and moderate earthquakes are common in this region. Normal faults along Narmada, Godavari and Koyna rifts are also reactivated occasionally.

3. Peninsular India : The Peninsular part of India consists of ancient rocks. Erosion has exposed the roots of the old mountains. The rocks are hard but softened by weathering. It is a geological stable region. Though, there are several faults but these are quire inactive. Very small earthquakes occur in this area which are mostly intraplate.

TABLE 1.9. Some Important Earthquakes in India

S. No.	Year	Name and place	Magnitude	Max. Intensity	Deaths
1.	1819	Kutch, Gujarat	8.3	VIII	>1500
2.	1897	Shillong	8.0	XII	>1500
3.	1905	Kangra, H.P.	8.6	X	20,000
4.	1934	Bihar, Nepal border	8.4	X	12,000
5.	1935	Quetta	7.6	X	30,000
6.	1950	Assam	8.5	X	1600
7.	1991	Uttarkashi	6.8	IX	800
8.	1993	Latur, Maharashtra	6.5	IX	8000
9.	1999	Chamoli	6.8	VIII	100
10.	2001	Bhuj	7.7	X	15000
11.	2004	Andman Island (Tsunami)	9.2	IX	2,50,000
12.	2005	Kashmir	7.6	VIII	86,000

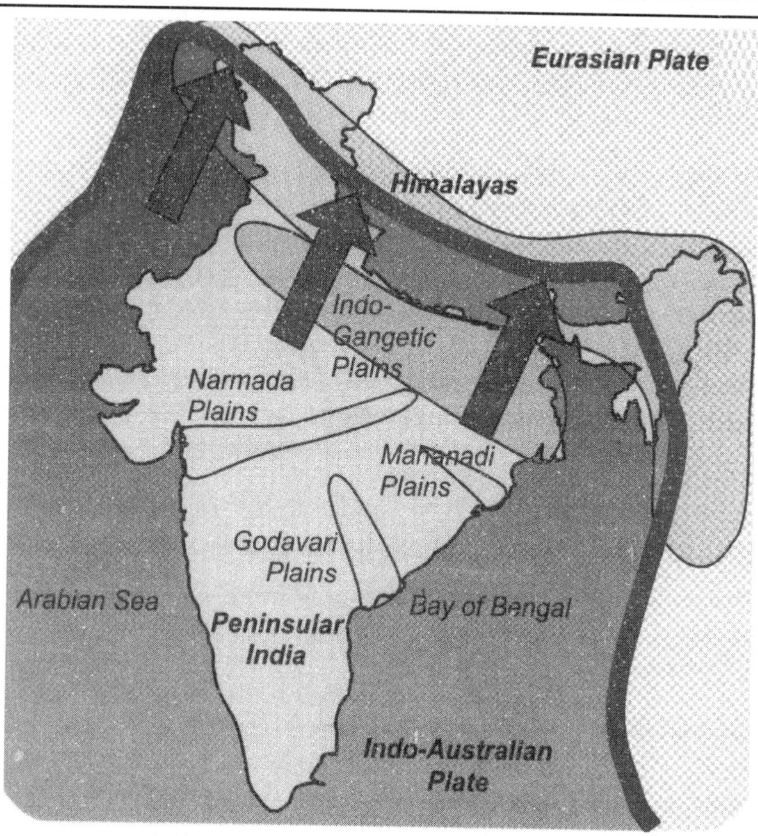

Fig. 1.15. Tectonic plates of India.

1.16. SEISMIC ZONES OF INDIA

The different geology at different locations in the country implies that the possibility of damaging earthquakes taking place at different locations is different. Thus a seismic zone map is required to identify these region. Based on the levels of intensities during past earthquakes, the 1970 zone map divided India into five zones I, II, III, IV and V (Fig. 1.16). The maximum modified Mercalli (MMI) intensity of shaking expected in these zones were V or less, VI, VII, VIII and IX and higher, respectively. Parts of Himalayan boundary in the north and north east and Kutch area in the west were classified as zone V.

The seismic zone maps are revised from time to time as more knowledge is gained about earthquakes, the tectonics and Seismic characteristics of the country. The Indian Standards published the first seismic zone map in 1962 which was later revised in 1967 and again in 1970. The map has been revised lastly in 2016 and it now has only four seismic zone—II, III, IV and V. (Fig. 1.15). The areas of zone I of the earlier 1970 map are merged with the seismic zone II. Also the zones in Peninsular region has been modified and Chennai now comes in Zone III (earlier it was in zone II).

The Indian seismic zone map

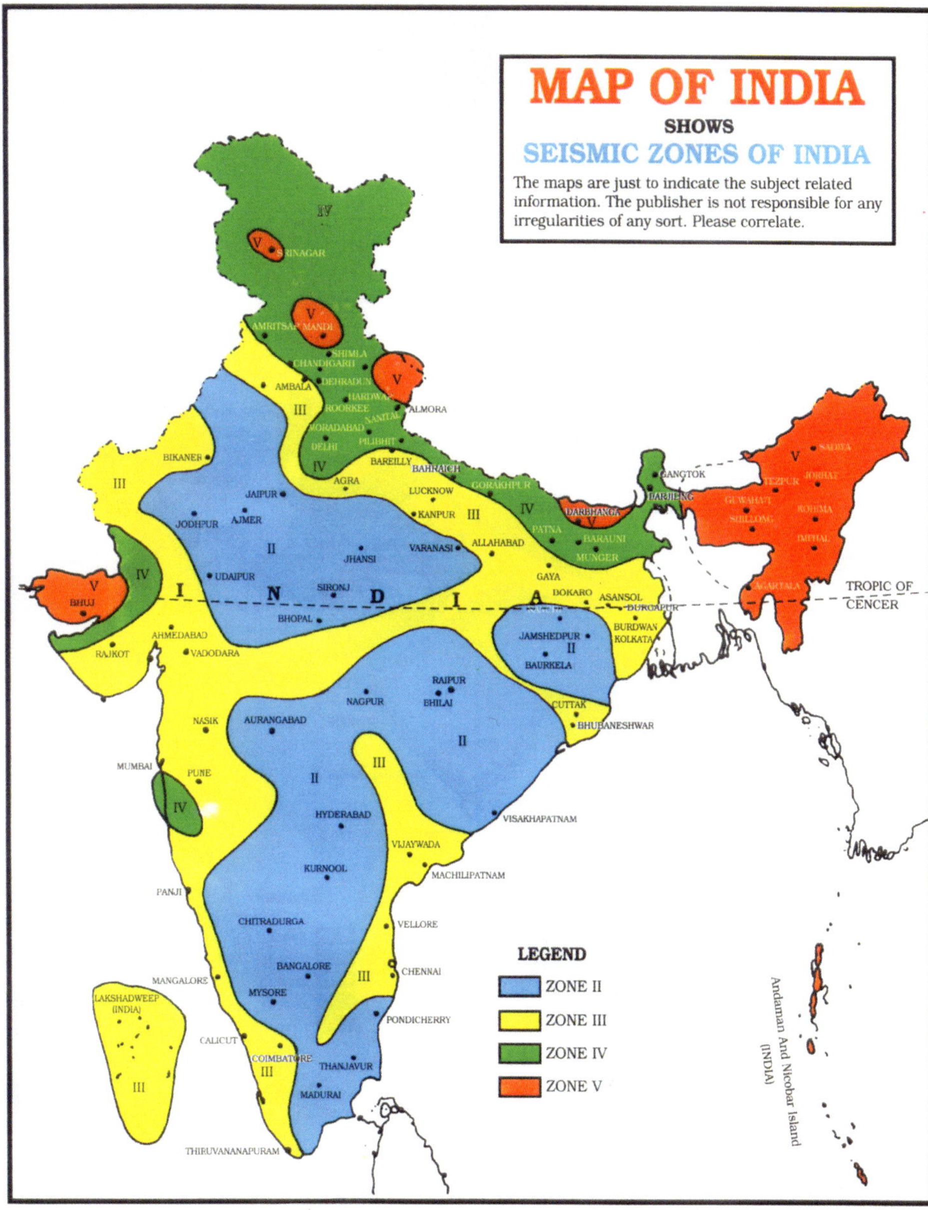

The Indian seismic zone map (Pre-revised) containing 5 Zones is as shown below :

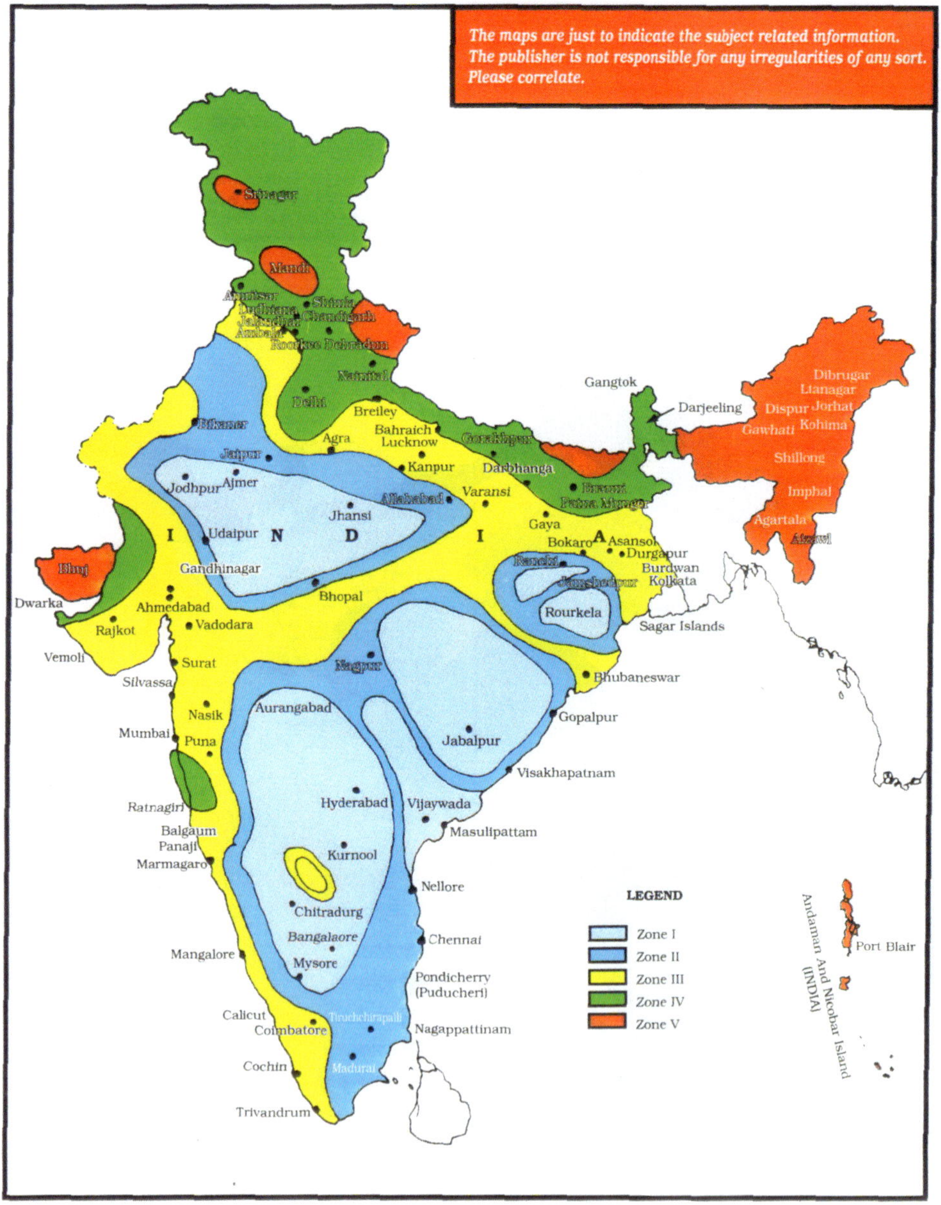

TERMS TO REMEMBER

1. **Earthquake :** The sudden shaking or movement of the earth's crust caused due to disturbances occuring inside the earth.
2. **Crust :** The uppermost part of the earth is called as crust.
3. **Tectonic features :** The nature of formation of rocks in the earth's crust featuring, dislocation, faults, folds, volcanoes etc. which are responsible for earth movement.
4. **Fault :** A fracture or a crack in the rocks (in the earth's crust), along which some definite movement has taken place.
5. **Hypocenter :** The point on the fault where slip starts and earthquake originates.
6. **Epicenter :** The point on the earth's surface directly above the hypocenter.
7. **Seismology :** The study of earthquakes and the seismic waves. It includes their origin, propagation and recording.
8. **Seismogram :** The record of the earthquake waves at the recording station.
9. **Accelerogram :** The variation of ground acceleration with time, recorded at a point on ground during earthquake.
10. **Earthquake Magnitude :** It is a number which is a measure of amount of energy released during an earthquake.
11. **Earthquake Intensity :** It is a measure of the effect of shaking at a place during an earthquake and is indicated by a Roman number.
12. **Isoseismal :** A countour joining places of equal earthquake intensity.
13. **Seismic Zoning :** Dividing a country into zones of similar seismic activities.
14. **Tsunami :** Tsunami means giant harbour waves or large sea waves caused by an earthquake activity at the ocean floor or near the coast.

EXERCISES

A. Objective Type Questions :

1. Study of earthquakes is called as
   (*a*) Seismography (*b*) Seismology
   (*c*) Quakeology (*d*) Plate tectonic
2. Inner most part of the earth is called as
   (*a*) Mantle (*b*) Core
   (*c*) Crust (*d*) Thrust
3. Which of the following is not a type of faults,
   (*a*) Deep slip faults (*b*) Strike slip fault
   (*c*) Oblique fault (*d*) Dip slip fault

4. Which of the following is not a type of Seismic wave.
   (*a*) P-waves (*b*) Lone waves
   (*c*) S-waves (*d*) Q-waves
5. Which seismic waves are the fastest.
   (*a*) P-waves (*b*) S-waves
   (*c*) Rayleigh waves (*d*) Love waves
6. The record of an earthquake is called as
   (*a*) Seismograph (*b*) Seismogram
   (*c*) Seismometer (*d*) Seismoscope
7. The point on the fault where slip starts is known as
   (*a*) Dip (*b*) Hypocentre
   (*c*) Epicentre (*d*) Strike
8. Which of the following scale is used to measure earthquake magnitude.
   (*a*) Richter scale (*b*) MMI scale
   (*c*) MSK scale (*d*) Energy scale
9. Intensity of an earthquake can be measured by
   (*a*) MSK scale (*b*) MMI scale
   (*c*) Richter scale (*d*) Both (*a*) and (*b*)
10. Seismic map of India given in IS1893 : 2002 divides the country into following number of zones
    (*a*) 5 (*b*) 4
    (*c*) 6 (*d*) 7
11. The uppermost part of the earth is called as
    (*a*) Core (*b*) Mantle
    (*c*) Crust (*d*) Mho
12. The amount of energy released during an earthquake is measured by
    (*a*) Magnitude (*b*) Intensity
    (*c*) Seismic waves (*d*) Epicentre
13. Isoseismals are contours joining places of equal
    (*a*) Magnitude (*b*) Intensity
    (*c*) Energy (*d*) Epicentre
14. Interplate earthquakes occur
    (*a*) Along the plates boundaries (*b*) Within a plate itself
    (*c*) Both (*a*) and (*b*) (*d*) None of the above.
15. Magnitude of an earth represents
    (*a*) Energy released (*b*) Acceleration of the ground motion
    (*c*) Damage caused (*d*) None of the above
16. Magnitude of earthquake is expressed as
    (*a*) Mercalli scale (*b*) Ritcher scale
    (*c*) MMI scale (*d*) MSK scale
17. Intensity of an earthquake
    (*a*) Increase with increase in epicentral distance
    (*b*) Decrease with increase in epicentral distance
    (*c*) Remains constant
    (*d*) None of the above

18. Relationship between energy released and magnitude of an earthquake as given by Ritcher is
(a) $\log_{10} M = 11.4 + 1.5E$
(b) $\log_{10} M = 11.4E + 1.5$
(c) $\log_{10} E = 11.4M + 1.5$
(d) $\log_{10} E = 11.4 + 1.5M$

19. Delhi lies in the seismic zone
(a) II
(b) III
(c) IV
(d) V

20. S-waves can travel through
(a) Solids
(b) Liquid
(c) Gas
(d) All of the above

B. Fill in the Blanks :

1. The sudden shaking of earth's crust is termed as ________.
2. The study of earthquake is known as________.
3. The uppermost part of the earth is called as________.
4. The point inside the earth at which an earthquake originates is called as________.
5. The focus of earthquake is called as________.
6. The point on the surface of the earth lying vertically above the focus is called________.
7. The depth of focus from the epicenter is termed as________.
8. The intensity of vibrations ________with the increase in epicentral distance.
9. The instrument which records ground vibrations is called as________.
10. For recording high intensity vibrations, special instruments called ________ are used.
11. ________are the imaginary lines joining points of equal seismic intensity.
12. ________means giant sea waves or harbour waves.
13. The minor shocks which precede and follow a major earthquake are called as ________and________.
14. A fracture or crack in the rocks of earth's crust along which some movement takes place is called as________.
15. As per IS 1893 : 2002, India is divided into ________seismic zones.
16. Delhi comes under seismic zone________.
17. The record of an earthquake is called as ________.
18. The variation of ground acceleration with time is called as________.
19. ________is dividing a country into zones of similar seismic activity.
20. ________waves are the fastest.
21. Most of the earthquakes that occur in the world are ________ .
22. ________waves can pass through solid only.
23. ________waves can pass through solid, liquids and gas.
24. S waves are also called as________.
25. ________ waves are most destructive in nature.
26. ________ of an earthquake is the measure of amount of energy released during an earthquake.
27. Earthquake size is defined in terms of its ________and________.
28. The magnitude of an earthquake is generally measured on ________scale.

29. _________ is a qualitative measure of the earthquake.
30. Hard rock foundations can vibrate more by _________and _________ waves.
31. Soft soils can vibrate more by _________waves and result into _________.
32. Epicentre of an earthquake can be determined from the _________.
33. Himalayan and Indo-Gangetic plains are in the _________seismic zone.
34. Modified version of Mercalli's scale is known as_________.
35. The earthquake with magnitude greater than 8.0 are known as_________.

C. State True/False :

1. We can prevent or predict earthquakes.
2. Focus is also known as epicenter
3. The intensity of vibrations decreases as the epicentral distance increases.
4. Seismograph is a graph showing ground motion.
5. Tsunami are a series of large sea waves caused by earthquake.
6. When two plates move towards each other, mountains are formed.
7. P waves are also called as shear waves.
8. S-waves are slower than P-waves.
9. Surface waves travel along the surface of the earth.
10. S-waves are surface waves.
11. Surface waves are most destructive in nature.
12. Magnitude is a qualitative measurement of earthquakes.
13. Increase by 1 unit in magnitude value of Richter scale leads to 31 times increase in amount of energy released.
14. M > 8.0 indicates major earthquakes.
15. Intensity value is generally based on the experiences of the people and performance of structures during earthquakes.
16. Interplate earthquakes occur within the plate itself.
17. Assam lies in the seismic zone V.
18. Magnitude of an earthquake is a fixed value.
19. Artificial disturbances, causing earthquakes are all man-made.
20. The amount of energy released during an earthquake is measured by its intensity.
21. Great boundary fault exists in Himalayan region.
22. Non-Tectonic, earthquakes are caused due to sudden slip in the fault.
23. The severity of ground shaking can be expressed in terms of peak ground acceleration.
24. The focal length of deep earthquake is greater then 500 K.M.
25. Seismic zoning is used for designing and planing earthquake resistant structures.

ANSWERS

A. Objective Type Questions :

1. (*b*)	2. (*b*)	3. (*a*)	4. (*d*)	5. (*a*)
6. (*b*)	7. (*b*)	8. (*a*)	9. (*d*)	10. (*b*)
11. (*c*)	12. (*a*)	13. (*b*)	14. (*a*)	15. (*a*)
16. (*b*)	17. (*b*)	18. (*d*)	19. (*c*)	20. (*a*)

B. Fill in the Blanks :

1. Earthquake
2. Seismology
3. Crust
4. Focus /Hypocenter
5. Hypocentre
6. Epicentre
7. Focal depth
8. decreases
9. Seismograph
10. Accelerographs
11. Isoseismal
12. Tsunami
13. Fore shocks and After shocks
14. Fault
15. four
16. IV
17. Seismogram
18. Acclerogram
19. Seismic zoning
20. P waves
21. Interplate
22. S waves
23. P-waves
24. Shear waves
25. Surface waves
26. Magnitude
27. Magnitude, Intensity
28. Ritcher
29. Intensity
30. P, S
31. surface, landslides
32. Seismogram
33. high
34. MMI
35. Great earthquake.

C. State True/False :

1. False	2. False	3. True	4. False	5. True
6. True	7. False	8. True	9. True	10. False
11. True	12. False	13. True	14. False	15. True
16. False	17. True	18. True	19. True	20. False
21. True	22. False	23. True	24. False	25. True

SHORT ANSWER TYPE QUESTIONS

1. Define the following terms :
   (*a*) Earthquake
   (*b*) Seismology
   (*c*) Epicentre
   (*d*) Hypocentre
   (*e*) Epicentral distance
   (*f*) Tsunamis
   (*g*) Foreshocks and after shocks
2. What are the three zones into which earth is divided ?
3. What are three major tectonic plates of India ?
4. Define surface waves and name their types.
5. Define earthquake size.
6. What is a seismogram ? Which instrument is used to record earthquake motion.
7. Define Isoseismal?
8. What do you understand by seismic zoning ?
9. What is a seismoscope ?
10. What do you mean by Accelerogram ?
11. What are seismic waves and name their types ?
12. What is the difference between interplate and intraplate earthquake ?
13. Define Isoseismal and explain their use.
14. What is the difference between magnitude and Intensity ?

15. Define seismology. Why is study of earthquake is important for Civil Engineers ?
16. What do you understand by dip and strike of the fault ?

LONG ANSWER TYPE QUESTIONS

1. Explain the structure of the earth.
2. Explain the causes of earthquake.
3. Define Earthquake size and give the classification of earthquakes.
4. What do you understand by Intensity of earthquake ? Which scales are used to measure intensity of earthquake.
5. What is a fault ? Explain different types of faults.
6. Explain the term seismic zoning. What is the use of seismic zoning maps ?
7. What are seismic waves ? Explain different types of seismic waves with the help of a diagram.
8. Write a short note on Tectonic features of India.
9. What do you understand by the term magnitude of an earthquake ? Give the classification of earthquake based on magnitude (Ritcher scale).
10. Give at least four differences between P and S waves.
11. With the help of a neat diagram, explain the working of a seismograph.
12. Explain the theory of plate tectonics.
13. In which Seismic zone do the following cities fall
    (*a*) Delhi
    (*b*) Bhuj
    (*c*) Chennai
    (*d*) Shimla
    (*e*) Guwahati
14. What is an Isoseismal ? What is seismic zoning and what is the use of seismic zoning maps ?
15. Define magnitude of an earthquake. Compare magnitude and intensity of an earthquake.

❑❑❑

2

CHAPTER

Structural Dynamics

INSIDE THIS CHAPTER

2.1. LOADS

The loads coming on the structures are basically of two types :

1. Static load
2. Dynamic load

1. Static load : Static loads are those loads which are not changing with time. For example Dead Loads, Imposed Loads of furnitures, surface finishings, snow load etc. The live loads of people and erection loads are also considered as static while analyzing structures.

2. Dynamic Loads : The loads which vary or changes with time are called as dynamic loads. For example wind loads, moving loads, machine vibrations, impact and blast loads etc. are dynamic loads. Dynamic loads are also of two types :

(*a*) *Periodic :* These loads are caused due to periodic nature of vibrations which means the motion repeats itself after equal interval of time. Rotating machine parts, vibration of pipeline structures caused by steady flow are some examples of periodic loads. The periodical motion can be resolved into sinusoidal forms (varying like a sine wave).

(*b*) *Non-Periodic or Random Loads :* The dynamic loads which don't follow any particular pattern over a period of time are called as non-periodic or random loads. Examples of non-periodic loads are earthquake loads, blast induced loads water waves impact loads etc. Sometimes, these loads are suddenly applied for a very short period of time and are called as transient loads.

In general, most of the civil engineering structures are analyzed by assuming that the applied loads are static. This is because

1. The structure is rarely subjected to dynamic loads.
2. The dynamic analysis is very complicated and time consuming.

But this tendency of ignoring the dynamic loads, may result in disaster sometimes. Thus they should be taken into account properly while analyzing and designing the structures.

2.2. DEGREES OF FREEDOM

A particle or a mass in space can undergo six movements : three translational and three rotational as shown in Fig. 2.1.

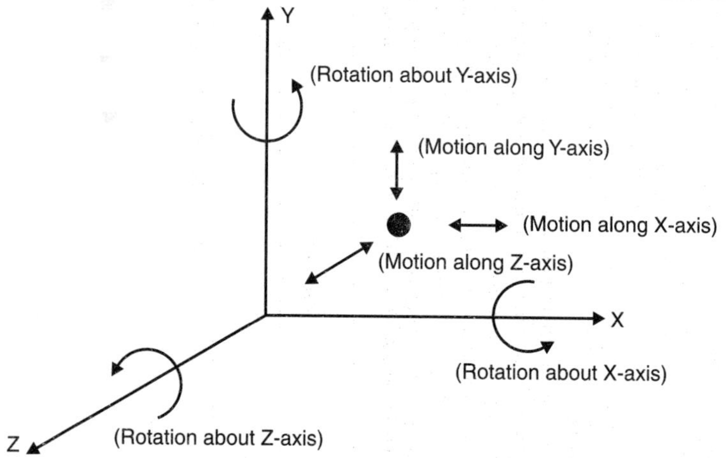

Fig. 2.1.

The three translational movements are :

1. Along X-direction
2. Along Y-direction
3. Along Z-direction

The three rotational movements are :

1. About X-axis
2. About Y-axis
3. About Z-axis

Thus a mass in space needs 6 independent coordinates to specify its configuration at any instant. Sometimes the mass may have freedom to move only in certain directions and the movement is constrained in other directions. Then the number of coordinates will be less than 6. *The number of coordinates required to describe the position of a vibrating mass at any instant is termed as Degrees of freedom.* The Figure 2.2 shows different systems of degrees of freedom.

Figure 2.2(*a*) shows a single mass fixed with a string (Simple Pendulum). This mass system is single degree freedom system as its motion is constrained in all directions except one *i.e.* (θ) rotation is possible only. Thus, at any instant the position of mass can be defined in terms of values of θ.

Figure 2.2(*b*) shows a two degrees of freedom system which comprises of a mass attached to two different springs. This system is free to move in two independent ways. It can move vertically along Y-direction and rotate about Z-direction.

Figure 2.2(*c*) shows a multi-mass system for such a system the total degrees of freedom will be the sum of degrees of freedom of each mass. If five masses such as the floors of a five storeyed building (considering the masses to be concentrated at floor levels) vibrate in one plane, five displacements for the five masses x_1 to x_5 will describe the deformed position of the structure. Thus, it will be termed as having five degrees of freedom.

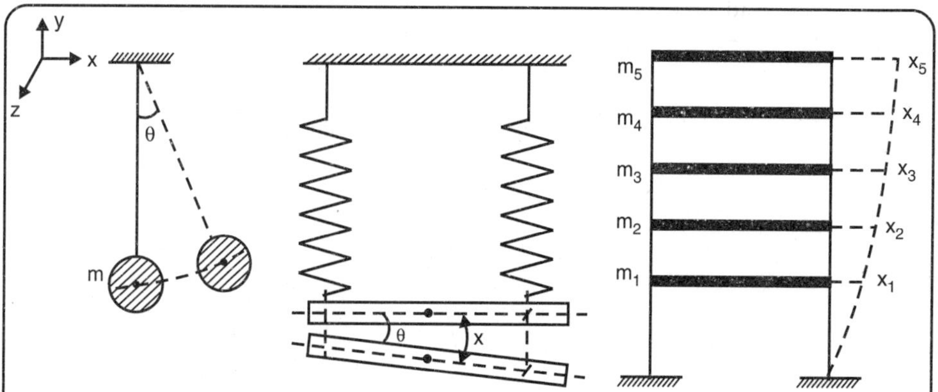

(*a*) Single degree of freedom. (*b*) Two degrees of freedom. (*c*) Five degrees of freedom.

Fig. 2.2. Degrees of freedom.

2.3. EQUILIBRIUM

When the resultant of a system of forces acting on a body is zero, the body is said to be in equilibrium. This concept of equilibrium is already studied by you in Statics. Let us consider a structure, acted upon by an external static force F. Under the action of force F, the structure undergoes a resultant displacement x then the static equilibrium equation for this structure is written as

$$F = K \cdot x$$

where $K.x$ is the restoring force developed in the structure which resists the deformation. K is the stiffness resistance of the structure and depends upon the type and material of the structure.

The same concept of static equilibrium (Resultant of forces = 0) is also applicable to dynamics. But before doing so let us study the various types of forces developing or acting on the structure under dynamic loading.

2.3.1. Dynamic Equilibrium

The Fig. 2.3 shows a mass attached to the top of a column and subjected to vibrations caused due to an externally applied force $F(t)$ which is varying with time. let us study the various forces acting on the mass.

The column is displaced from its original position and the displacement of mass m is 'x' at any time t. $\dot{x}$ is velocity and $\ddot{x}$ is the acceleration of the mass at time t.

It is clear that the column will tend to bring the mass back to its original position. Thus, a **restoring force (F_R)** is exerted by the column which is equal to K times x. K is the stiffness of the column. The mass will come back and move to the other side and vibrate. This vibration will continue forever. But in real, friction with air, friction between the particles of the system or connections and junctions etc. opposes the motion and hence, after sometime, the vibrations will die out. These forces which opposes motion are called as **damping forces** and depend on the materials, nature of joints, quality of construction and type of foundation etc. The damping force can be determined experimentally only. It is generally taken as proportional to the velocity of the mass system and called as viscous damping. Figure 2.4 shows the effect of damping on vibrations.

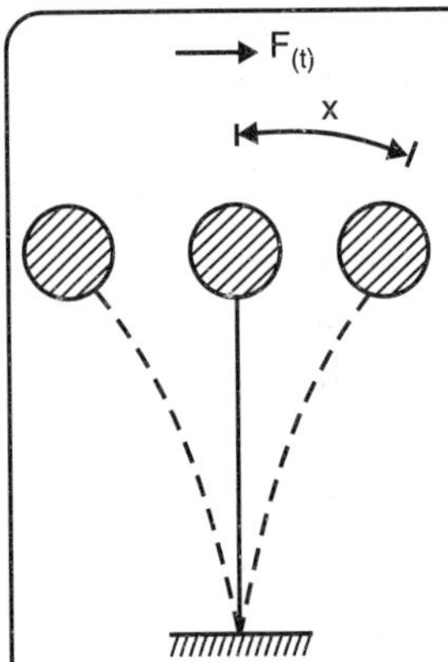

Fig. 2.3.

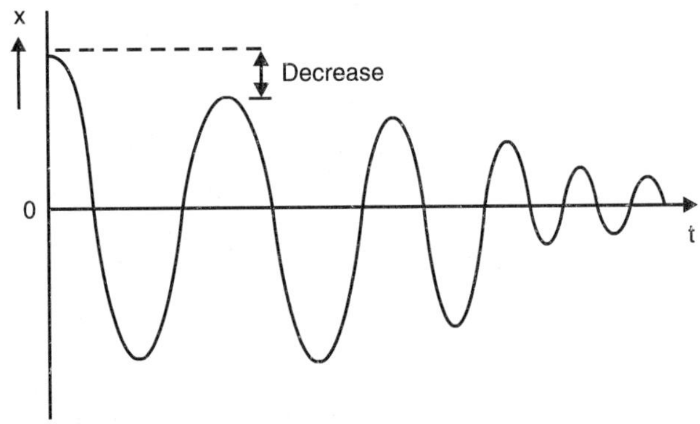

Fig. 2.4. Damped free vibrations.

The various forces acting on the mass under dynamic loading are as follows :

(1) Restoring force, $F_R = Kx$ where K is the stiffness and this force acts opposite to the direction of motion.

(2) Damping force, $F_D = C\dot{x}$, C is the coefficient of damping and $\dot{x}$ is the velocity of the mass system, acts in the direction opposite to direction of motion.

(3) $F_{(t)}$ = External force.

By Newton's Second law of motion,

$$F_{(t)} - F_R - F_D = \text{Rate of change of momentum}$$

$$= \frac{d(mV)}{dt}$$

In civil engg. structures mass m is constant and hence

$$F_{(t)} - F_R - F_D = m \cdot \frac{dv}{dt} = m\ddot{x}$$

$$F_{(t)} - Kx - C\dot{x} = m\ddot{x}$$

$$\boxed{F_{(t)} = m\ddot{x} + C\dot{x} + Kx} \qquad ...(i)$$

The equation (i) is called as equation of dynamic equilibrium.

$m\ddot{x}$ or rate of change of momentum is considered to have the same effect as that of applying a fictitious force F_i, opposite to the direction of motion and equal to $m\ddot{x}$. This force F_i is called as Inertia Force.

2.4. EFFECT OF EARTHQUAKE MOTION ON STRUCTURES

Earthquake causes shaking of the ground. The structures resting on the ground will be subjected to motion at base. Thus, the base of the structure moves but the structure has a tendency to stay in its original position due to inertia. The the structure can be supposed to be acted upon by a force, called Inertia force (which resists the change of the state of the structure). For example in a building whose roof is supported on columns [Fig. 2.5(*a*)] and the columns are fixed to the foundation (base), when the ground moves the foundation alongwith the columns tries to move the roof. But roof has a tendency to stay back, hence Inertia force equal to mass of roof times acceleration acts on the roof in the direction opposite to the ground motion [Fig. 2.5(*b*)].

During earthquake motion, the columns of the building are subjected to displacements. But, they also have a tendency to come back to the straight position (original position). Thus, restoring forces, which depends upon the stiffness of the columns and the displacement (x) are developed ($F_D = Kx$). These act also opposite to the direction of motion. It is clear that if the columns are stiffer than the restoring forces are more.

As already explained, every structure possess some damping which depends upon its materials connections and components etc. Thus damping forces equal to $C\dot{x}$ (C = damping coefficient and $\dot{x}$ is the velocity of ground motion) are

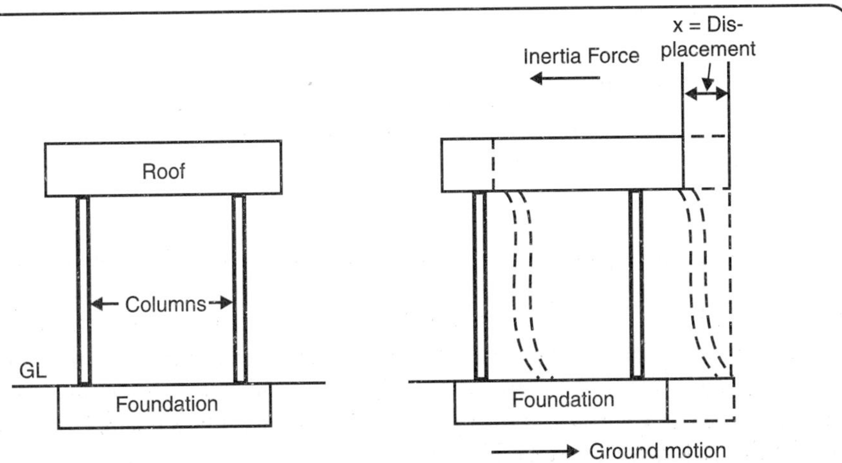

Fig. 2.5. Inertia force.

developed which also act opposite to the direction of motion. Thus, the summation of forces acting on the building is

$$F_{(t)} = m\ddot{x} + C\dot{x} + Kx$$

which is same as equation for dynamic equilibrium.

Thus, the equation of motion (dynamic equlibrium) is applicable to buildings under earthquake motion also. Here $F_{(t)}$ is the dynamic force due to earthquake motion.

2.5. FUNDAMENTAL NATURAL PERIOD

Let us study the behaviour of a building during earthquake shaking. When the ground shakes, the base of a building moves with the ground and the building also vibrates backward and forward (Oscillations). If the building is flexible there will be larger vibrations as compared to stiffer buildings. As already explained, every structure possess some damping which depends on its components, materials and connections. This damping resists the vibrations of the structure. Therefore after sometime the building will come back to its original position (*i.e.* the oscillations will die out). These oscillations are periodic which means the motion repeats itself after equal interval of time (Fig. 2.6). *The time taken in second for completing one cycle of oscillations (Fig. 2.7) is called Fundamental natural period (T) of the building.*

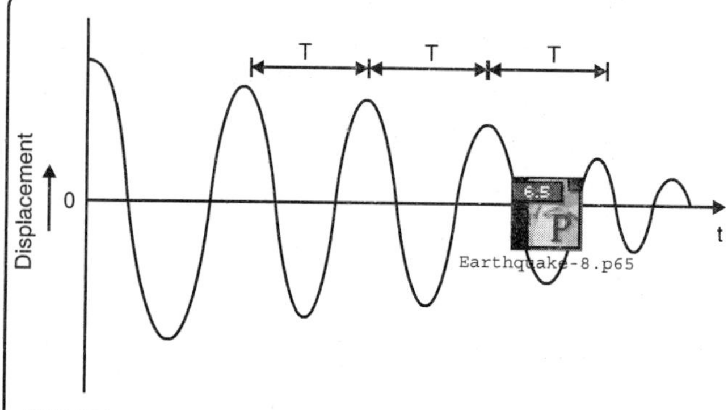

Fig. 2.6. Periodic motion.

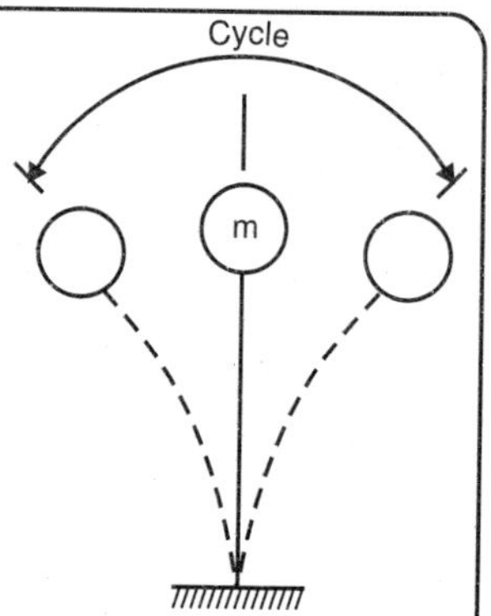

Fig. 2.7.

The value of T depends on the following :

1. Building flexibility 2. Mass

1. Building flexibility : If the building is flexible then movement is larger that means the time taken for completing one cycle of oscillations will be more and hence fundamental natural period (T) will be higher.

2. Mass : If the mass is more then longer is fundamental natural period (T).

In general, taller buildings are more flexible and have larger mass and therefore a larger Fundamental period (T). On the other hand, low to medium rise buildings have smaller T.

On the basis of T, building may be classified as :

1. Rigid buildings : The buildings whose fundamental period is less than 0.3 seconds.

2. Semi rigid : The buildings whose fundamental period is between 0.3 to 1.0 seconds.

3. Flexible buildings : The buildings whose fundamental period is greater than 1.0 seconds.

Fundamental natural period is a property of the building. Some example of natural periods of different types of structures are given in Table 2.1.

TABLE 2.1.

Type of Building	Fundamental Natural Period (Sec.)
1-2 Storeyed building	0.05 sec
2-6 Storeyed building	0.5 sec
Multistoreyed building (≥ 15 storeys)	1 sec
Gravity dam	0.8 sec
Suspension bridge	6 sec
RCC chimney	2 sec
Overhead water tank	4 sec

TERMS TO REMEMBER

1. **Static Load :** The loads which are constant with time.
2. **Dynamic Load :** The loads which changes with time. They may be periodic or random in nature.
3. **Degrees of Freedom :** The number of coordinates required to describe the position of vibrating mass at any instant of time.
4. **Damping :** The effect of internal friction, material slipping, sliding etc. which reduces the amplitude of vibration.
5. **Inertia :** The tendency of a body to resist the change of state is called as Inertia.
6. **Inertia Force :** A fictitious force supposed to act on a body under dynamic equilibrium which is opposite to the direction of motion. Its magnitude is equal to mass multiplied by acceleration.

EXERCISES

A. Objective Type Questions :

1. Inertia force is equal to
   (*a*) Mass × Velocity (*b*) Mass × Acceleration
   (*c*) Mass × Displacement (*d*) None of these
2. The loads which vary with time are called as
   (*a*) Static loads (*b*) Imposed loads
   (*c*) Dynamic loads (*d*) Live loads
3. A mass in space has following number of degrees of freedom.
   (*a*) 4 (*b*) 8
   (*c*) 5 (*d*) 6
4. Damping has the following effect on vibrations
   (*a*) To decrease the amplitude of vibrations

(*b*) To increase the amplitude of vibrations
(*c*) To increase the time period
(*d*) No effect

5. Restoring force is equal to
(*a*) Mass × Acceleration (*b*) Mass × Displacement
(*c*) Stiffness × Displacement (*d*) None of the above

6. Equation of dynamic equilibrium is written as
(*a*) $m\dot{x} + C\ddot{x} + Kx$ (*b*) $m\ddot{x} + C\dot{x} + Kx$
(*c*) $m\ddot{x} + K\dot{x} + Cx$ (*d*) None of the above.

7. If the roof of a building is heavy, then the inertia force will be
(*a*) More (*b*) Less
(*c*) Not effected

8. If the building is flexible then fundamental natural period of building will be
(*a*) Shorter (*b*) Longer
(*c*) Not effected

9. If the building has more mass the fundamental natural period will be
(*a*) Longer (*b*) Shorter
(*c*) Not affected

10. Earthquake load is an example of
(a) Static load (*b*) Line load
(*c*) Non-periodic dynamic load (*d*) Periodic dynamic load

B. Fill in the Blanks :

1. The loads which don't change with time are called as ________.
2. Earthquake load is an example of ________ load.
3. A point in space has ________ degrees of freedom.
4. The loads which are applied suddenly for a very short period of time are called as ________.
5. The equation of static equilibrium is written as F = ________.
6. The ________ force tends to bring the structure back to its original position.
7. The frictional forces which oppose motion are called as ________.
8. The equation of dynamic equilibrium is F(t) = ________,
9. The force $m\ddot{x}$ is called as ________.
10. The time taken in seconds for completing one cycle of oscillation is called as ________ of the structure.
11. The fundamental natural period depends upon ________ and ________.
12. The tendency of a body to resist change of state is called as ________.
13. The fundamental time period of high rise buildings (15 storeys) is ________.
14. The fundamental time period of ________ buildings is less than 0.3 seconds.
15. ________ is proportional to the velocity of the mass system.

C. State True/False :

1. The live load of people is treated as dynamic loads.
2. The loads which vary with time are called as dynamic loads.
3. Simple pendulum has only one degree of freedom.
4. The inertia force acts in the direction of ground motion.
5. The inertia force is equal to $m\ddot{x}$.
6. F = Kx represents the equation of dynamic equilibrium.
7. The damping force acts in the direction opposite to that of ground motion.
8. A taller building will have lower fundamental natural period.
9. A heavier building will have higher fundamental natural period.
10. The inertia force is an imaginary force.
11. The damping force tries to bring the mass back to its original position.
12. Damping force depends on the height of structure.
13. Stiffer columns means greater restoring force.
14. Periodic loads are a type of static load.

ANSWERS

A. Objective Type Questions :

1. (*b*)	2. (*c*)	3. (*d*)	4. (*a*)	5. (*c*)
6. (*b*)	7. (*a*)	8. (*b*)	9. (*a*)	10. (*c*)

B. Fill in the Blanks :

1. static loads	2. dynamic	3. 6
4. transient loads	5. Kx	6. restoring
7. damping forces	8. $m\ddot{x} + c\dot{x} + kx$	
9. inertia force	10. fundamental natural period	
11. building flexibility, mass	12. Inertia	13. 1.0
14. rigid	15. damping force	

C. State True/False :

1. False	2. True	3. True	4. False	5. True
6. False	7. True	8. False	9. True	10. True
11. False	12. False	13. True	14. False	

SHORT ANSWER TYPE QUESTIONS

1. Define the following terms :
   (*i*) Static load
   (*ii*) Dynamic loads
   (*iii*) Degree of freedom
   (*iv*) Inertia force
2. What do you understand by the term fundamental natural period ?
3. What is the difference between static and dynamic loads ? Given examples of each.

4. Write the equation of dynamic equilibrium and explain the various notations used in it.
5. How many degree of freedom will the following structures have :
   (*i*) A mass in space
   (*ii*) Simple pendulum
   (*iii*) A beam supported on springs.

LONG ANSWER TYPE QUESTIONS

1. Derive the equation of dynamic equilibrium for a building subjected to earthquate motion.
2. What is the fundamental natural period ? Explain the factors on which it depends.
3. Explain the following with reference to the equation of dynamic equilibrium.
   (*i*) Inertia force
   (*ii*) Restoring force
   (*iii*) Damping force
4. Explain the term degree of freedom with the help of some examples.
5. What is inertia force ? Explain the affect of earthquake motion on a building ?
6. Define degrees of freedom. Give example of single, two and multiple of freedom.
7. Explain different types of loads coming on the structures. Give examples of each kind.
8. What is the difference between static loads and dynamic load ? Explain with the help of examples.
9. Derive the equation of dynamic equilibrium for a simple mass column system.
10. Explain the various forces developed in a building during earthquake.

❑❑❑

3

CHAPTER

Seismic Performance of Traditionally Built Constructions

INSIDE THIS CHAPTER

3.1. INTRODUCTION

More than 80% of India's population still lives in small towns and villages. They use traditional methods and materials of construction. The materials used in these constructions are generally locally available like bricks, stone, earth, wood etc. These are generally masonry constructions or sometimes composite with concrete or steel. Masonry is commonly used for housing not only in India but also in many other countries of the world. The masonry has got many advantages which are discussed later in this chapter but its main disadvantage is its low earthquake resistance. Studies have shown that masonry buildings have suffered maximum damage during earthquakes. The collapse of these buildings result in heavy loss of life. This loss of lives and damage can be minimzed by studying the various types of damages observed in traditional construction and then causes. Thus, this chapter focusses on the various earthquake effects, types of traditionally built constructions in India, their advantages and disadvantages. The chapter also explains the behaviour of masonry construction under earthquake loading, various modes of failure of masonry and the commonly observed damages and their causes. All these studies have resulted in development of various methods for improving the earthquake resistance of the traditionally built construction, which will be discussed in next chapter.

3.2. EARTHQUAKE EFFECTS

There are four basic causes of earthquake induced damage of structures.

(1) Ground Shaking : The main cause of earthquake induced damage is ground shaking. As the earth vibrates, all buildings on the ground surface will be subjected to vibrations. Earthquake induced accelerations, velocities and displacements can damage or destroy a building unless it has been desgined and constructed properly.

(2) Ground Failure : Earthquake induced ground failure is observed in the form of ground rupture along the fault zone, landslides, settlement and soil liquefaction (soil behaves like a liquid). Ground rupture means cracking of ground. It may be very small or may extend up to hundreds of kilometers. While landslide can destroy a building completely the settlement may only damage the building.

Soil liquifaction generally occur in low density saturated sands. In this the soil flows and behaves like a liquid as a result of earthquake shaking. It is important to consider liquefaction while designing dams, bridges, and buildings on soft/loose soils.

(3) Tsunamis : Tsunamis or large sea waves are produced as a result of sudden movement of ocean floor. As the water waves approach land, their velocity decreases and their height increases (sometimes upto 10 m or more !) Tsunamis can be dangerous for buildings built in coastal areas. On Dec. 26, 2004 a deadliest Tsunami (Indian ocean) hit Sumatra, India, Srilaka resulted in more than 2,20,000 deaths.

(4) Fire : When the fire starts after an earthquake (electrical short circuit or from kitchen), it is difficult to extinguish it, because earthquake results in loss of water supply, breaking of pipes, traffic jams and panic. For example in 1923 Tokyo earthquake, 50% of the total number of houses were burnt due to fire and many people were killed.

3.3. TRADITIONALLY BUILT CONSTRUCTION IN INDIA

In India, masonary constructions are generally done by using locally available materials like

1. Stone
2. Brick
3. Adobe (sun-dried clay brick)
4. Timber
5. Mud
6. Concrete blocks

These materials are used along with different types of mortars (to bind them). In India, lime, mud, cement and sand are generally used in fixed proportions as mortar. The properties of masonry constructions depend basically on the masonry unit and the type of mortar used. The common construction practices generaly depend upon the topography of the area and the locally available material. For example, in plains, masonry is generally made of burnt clay bricks and cement mortar. In rural areas earthen buildings are very common. In some villages composite construction of earth with wood, bamboo and cane is also prevalent. In hilly areas like Uttarakhand, Himachal Pradesh and Rural Maharashtra, Masonry is generally made of stone with lime/mud or cement mortar. The stone walls are generally very thick (600–1500 mm). The roofing system, generally consists of wooden logs (rafters) supporting a large thickness of earth. In rural areas thatch and asbestos cement sheets are also used as roof coverings. In hilly areas pitched (sloping) roofs are very common. These are made of wooden joists or wooden planks supporting the tiles.

Japan Earthquake and Tsunami in 2011

Now a days reinforced cement concrete is also used with brick masonry walls in urban and semiurban areas. These buildings are constructed using isolated R.C.C. columns, R.C.C. slabs and beams resting on the masonry walls without proper design. All these types of constructions are dealt one by one later in this chapter.

3.4. ADVANTAGES AND DISADVANTAGES OF MASONRY CONSTRUCTION

The use of masonry buildings in India and many parts of the world is due to its many advantages over other types of constructions. Some of the advantages of masonry constructions are listed below :

(*a*) **Use of locally available materials :** The traditional masonry construction uses local materials and masons. Thus, the availability is easy and transportation cost is reduced.

(*b*) **Need of less skilled labour :** The masonry constructions require less skilled labour as they don't use any professional technique and the construction is simple.

(*c*) **Easy and inexpensive repair :** The repair is also easy and cheap due to cheaper material and local masons.

(*d*) **Less formwork :** This type of construction requires very little formwork and hence speedy construction.

(*e*) **Thermal Comfort and Sound Control :** The masonry constructions are generally massive and provide good insulation against heat and sound.

(*f*) **Possibility of easy alteration after construction**

All the above factors lead to overall economy and ease in the masonry construction and making it a favourite for all regions.

Despite being favourite material of construction, masonry has following disadvantages :

(*a*) Masonry structures are very heavy and in case of collapse, they result in heavy damage and loss of lives.

(*b*) Masonry has very low tensile strength and low shear strength, especially with poor mortars.

(*c*) Masonry show brittle behaviour in tension as well as compression.

(*d*) The wall to wall connection and roof to wall connection is generally weak.

(*e*) In masonry constructions, stress concentration occurs at corners of windows and doors.

(*f*) Defects in constructions, such as use of substandard materials, unfilled joints between masonry units (brick, stone etc.), non-plumb walls, improper bonding between walls etc. are very common in masonry construction.

3.5. BEHAVIOUR OF MASONRY CONSTRUCTION DURING EARTHQUAKES

Figure 3.1 shows a masonry building. A masonry building has three basic components :

(1) Roof : Masonry buildings, generally have heavy roofs which results in the higher earthquake forces (inertia forces)

(2) Walls : These are mostly load bearing. The walls transfer the vertical load and inertia forces from the roof to the foundation.

(3) Foundation : It is the part below the ground level which transfer all the forces to soil safely.

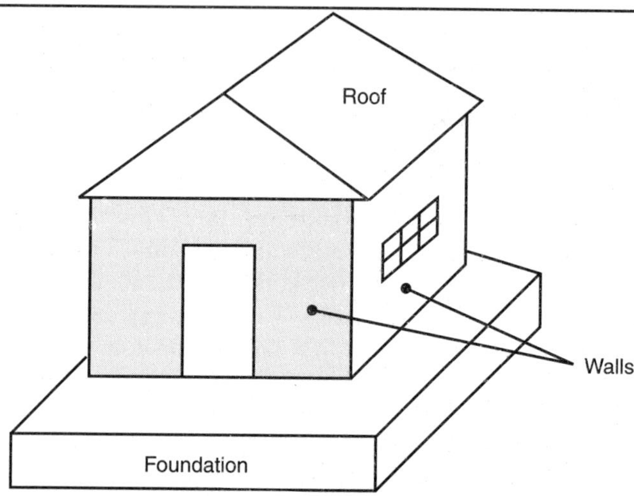

(a) Basic components of a masonry building.

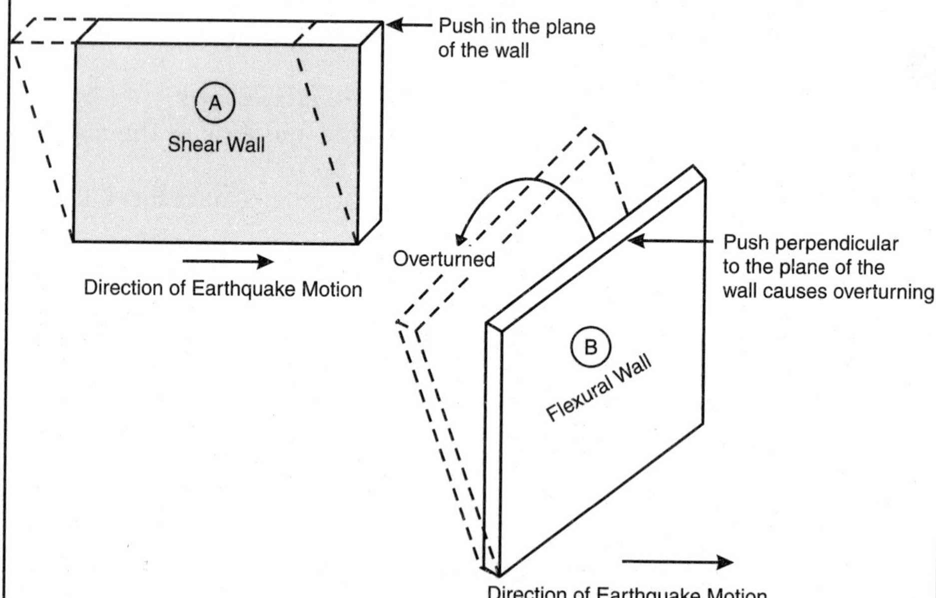

(b) Shear walls and flexural walls.

Fig. 3.1.

During an earthquake, the ground shakes in all the three directions but the horizontal vibrations are the most damaging to masonry buildings. These horizontal vibrations result in large horizontal Inertia forces (Chapter 2). These horizontal forces are transferred to the walls or columns from the roof and then to the foundation and finally to the soil. The inertia forces act opposite to the direction of earthquake motion and are reversible in nature. Thus at any instant, some walls of the building are parallel and some walls are transverse (perpendicular) to the direction of ground motion caused due to earthquake. For example in Fig. 3.1(*a*) walls A are parallel to the direction of earthquake shaking and walls B are perpendicular to the direction of earthquake shaking for the assumed direction of earthquake motion.

Let's study the behaviour of these walls. The horizontal inertia force act transverse to the wall B and hence tries to push it horizontally at the top in a direction perpendicular to its plane. In other words, the horizontal force tends to overturn or bend the wall out of its plane [Fig. 3.1(*b*)]. The walls which are transverse to the direction of the motion are called as **flexural walls**. The wall A is parallel to the direction of earthquake shaking. Hence the wall A will experience the inertia force which tries to push (slide) it in its plane [Fig. 3.1(b)]. In other words the wall will be subjected to shear and bending in its plane. These walls (A) are called as **shear walls**. As soon as the direction of earthquake motion changes, the walls behaviour will also change accordingly. The performance of the walls and other components of a masonry building under earthquakes can be understood by studying the action of earthquake forces on them and their failure mechanisms.

3.6. FAILURE MECHANISM OF A MASONRY BUILDING

A masonry building can fail in various ways under the action of earthquake forces. Some of the common modes of failure are explained below :

(1) Out of plane failure
(2) In plane failure
(3) Connection failure
(4) Diaphragm failure
(5) Failure due to openings in the walls
(6) Pounding
(7) Non-structural component failure.

3.6.1. Out of Plane Failure

Let's again consider the wall B shown in Fig. 3.1(*a*). The wall is transverse to the direction of earthquake motion. Thus it is subjected to Inertia Forces (earthquake forces) perpendicular to its plane. These forces tend to overturn or bend the wall as shown in Fig. 3.1(*b*). The wall has very less resistance in this direction due to small depth. Thus, the wall will collapse easily due to overturning. The bending of the wall results in development of tensile stresses. Masonry is very weak in tension and hence it cracks. This cracking can lead to full or partial collapse of the wall. This type of failure is called as out of plane failure. This type of failure features :

(1) Vertical cracks in the corner of walls and at T junctions.
(2) Cracks at lintel and top of piers
(3) Cracks at the level of roof
(4) Peeling of masonry
(5) Collapse (full or partial) of the exterior wall
(6) Horizontal cracks along the facade.

The tendency of the walls to overturn when pushed in the perpendicular direction is increased with the length of the wall. Long unsupported length means easier collapse (Fig. 3.2). Similarly tall and slender walls are easily over turned during earthquake motion. Thus the tendency of a wall to overturn can be reduced by limiting its length to thickness and height to thickness ratio.

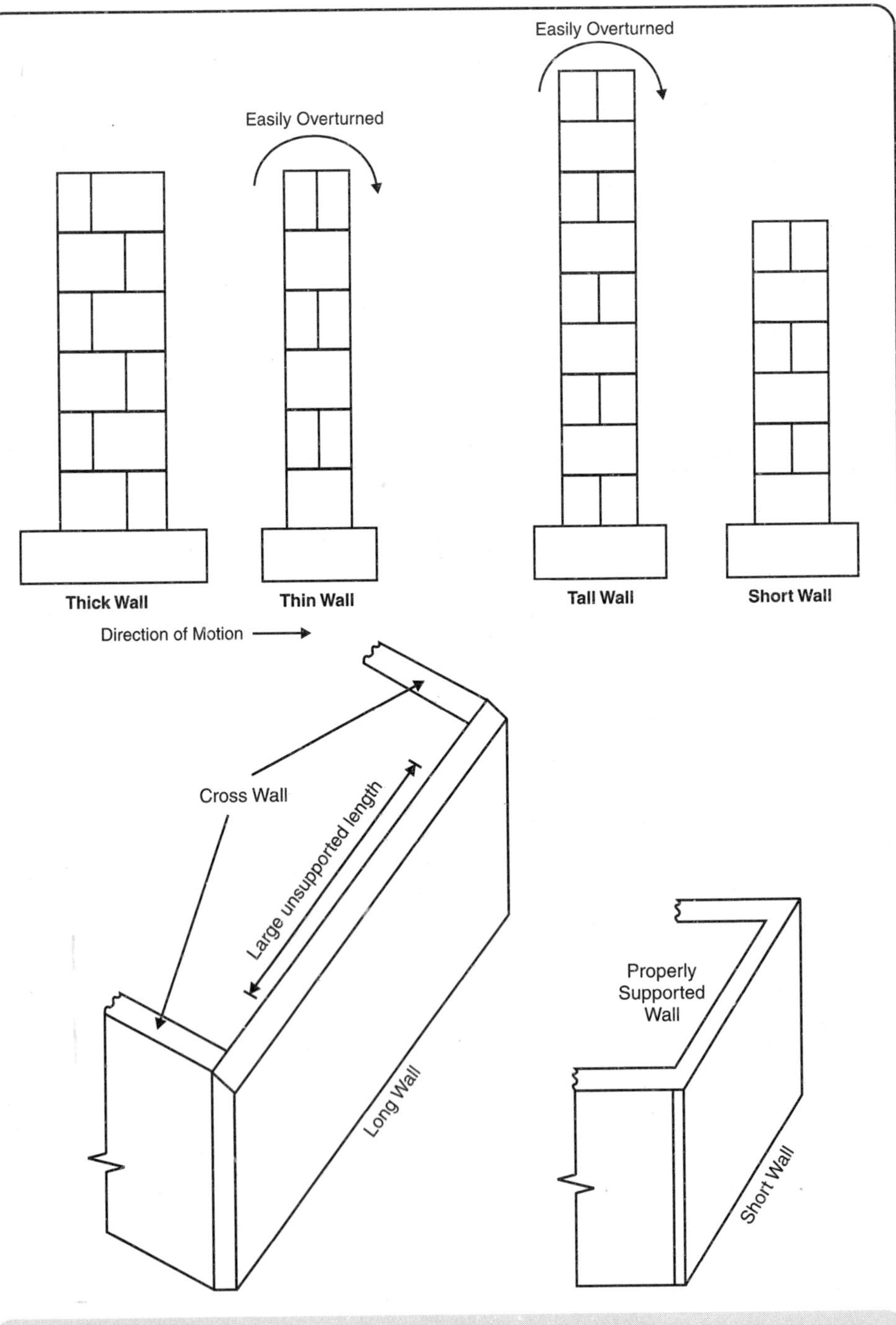

Fig. 3.2.

3.6.2. In Plane Failure

Let's consider the wall A of Fig. 3.1(*a*). This wall is subjected to ground motion in its plane. Thus it is subjected to inertia forces which causes shear (push or slide) and bending in the plane of the wall. This wall offers good resistance because of large depth in the plane of the bending. The damage is more if the wall is long and slender. A wall with small length to width ratio, generally develop horizontal cracks due to tension caused by bending while a wall with moderate length to width ratio shows shear failure by diagonal tension cracks. A wall with large length to width ratio may develop diagonal tension cracks at both sides because of shear and horizontal cracks at the middle due to bending [Fig. 3.3(*a*, *b*, and *c*)]. The other features of In plane failure are following :

(1) Diagonal tension cracks (×) in short piers between opening.

(2) Vertical cracks at walls intersections.

(3) Spalling of material.

(4) Crushing of corners.

(5) Separation and expulsion of the corner of two walls.

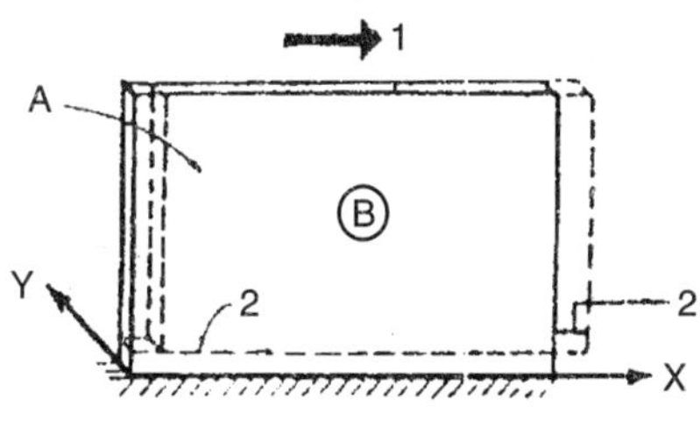

(a) Unreinforced wall B with small length-to-width ratio

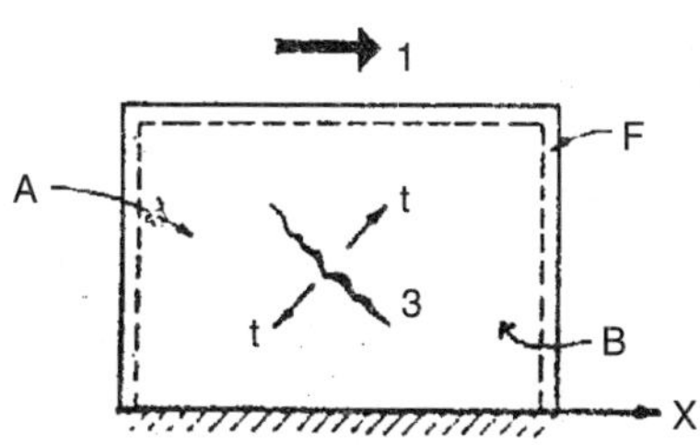

(*b*) Wall B with moderate length-to -width ratio

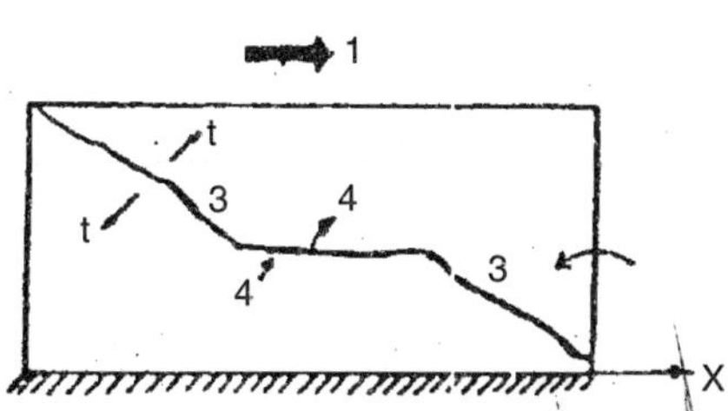

(*c*) Wall B with larger length-to-width ratio

1 = Earthquake force, 2 - Sliding, 3 = Diagonal cracking, 4 = Horizontal cracking, t = Diagonal tension

Fig. 3.3. Failure mechanism of shear wall.

3.6.3. Connection Failure

Now consider the walls A and B together as shown in Fig. 3.4. For the given direction of earthquake shaking, walls A act as shear walls and walls B act as flexural walls. Walls B are subjected to overturning (out of plane bending) by the horizontal inertia force and walls A are subjected to shear and bending in the plane of wall. If the walls are tied together properly (bonded together) then they act like a box as shown in Fig. 3.4(*b*). The walls B can take advantage of good resistance of wall A during earthquake shaking. The walls A offer resistance against inertia force in their plane but also support walls B from overturning. Thus, the box action provides greater resistance to the horizontal loads. But if the walls are not tied together properly then walls B will overturn and may cause collapse of the building as a whole. This type of failure is called as

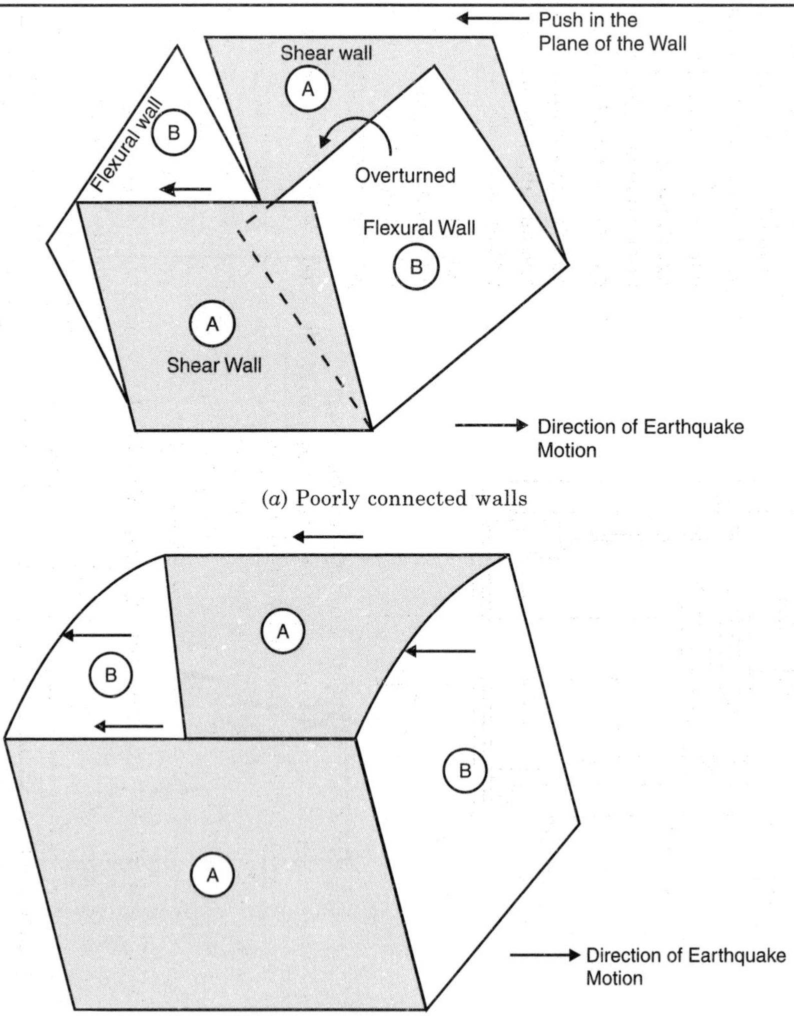

(*a*) Poorly connected walls

(*b*) Properly connected Walls resulting in box-action

Fig. 3.4. Connection failure.

connection failure [Fig. 3.4(*a*)] This type of failure is commonly seen in the form of diagonal cracks on masonry walls and collapse of corner materials.

The simplest way to make masonry building stronger to resist earthquakes is to make them properly tied or bonded so that they act as one unit (like a box).

3.6.4. Diaphragm Failure

Consider a building with a roof and the walls (A and B) subjected to earthquake. The roof is subjected to earthquake force as shown in Fig. 3.5. The roof/slab will transfer the earthquake force (inertia force) to the walls, causing shearing and bending in them. To transfer the forces the roof must have enough strength in bending in the horizontal plane. This action is called as **diaphragm** action. Reinforced concrete or reinforced brick slabs have good strength and act as **rigid diaphragms**. But, other types of roofs or floors made of timber or joists with brick tile coverings are very flexible and are called as **flexible diaphragms**. If the roof is rigid and acts as horizontal diaphragm, the inertia forces will be distributed to the walls in proportion to their stiffness. For example, wall A (In plane) is much stiffer than wall B (out of plane). Thus, wall A will get larger share of inertia force and hence the bending (overturning) of wall B will be less. On the other hand, if roof is flexible, the inertia forces will be distributed equally to the walls on which it is supported. Thus, the wall B will be subjected to greater inertia force and tend to overturn. This type of failure is called as diaphragm failure. *In this failures cracks develop near the corners of the wall and may lead to separation of the walls and in turn collapse of the building.*

The roofs and floors which are rigid and flat and are tied (bonded) to the walls properly do not show any sign of diaphragm failure. The roofs, which simply rest on the walls, will offer resistance to earthquake shaking by friction only. If the intensity of inertia force is greater than the frictional resistance, then the roof may slide off the walls. This type of failure is common in the roofs consisting of timber joists with brick tiles over them and the joists not properly tied. The sliding of tiles can cause damage and injury to people.

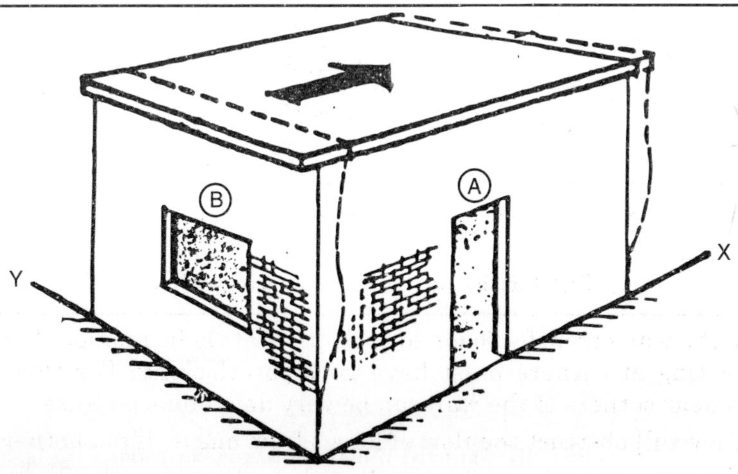

Fig. 3.5. Diaphragm failure.

3.6.5. Failure due to Openings in Walls

Openings are necessary in a building. But the location and size of the openings in walls affect the performance of masonry building in earthquake. Let's consider a four wall system of single storey masonry building (Fig. 3.6). During earthquake shaking inertia forces act on the walls. Some walls (A) will have in plane bending and some walls will have out of plane bending (B). The walls B will take support from the wall A for the direction of shaking shown in Fig. 3.6. If the direction of shaking changes and become perpendicular to shown in Fig. 3.6 then the action of walls B and A will reverse.

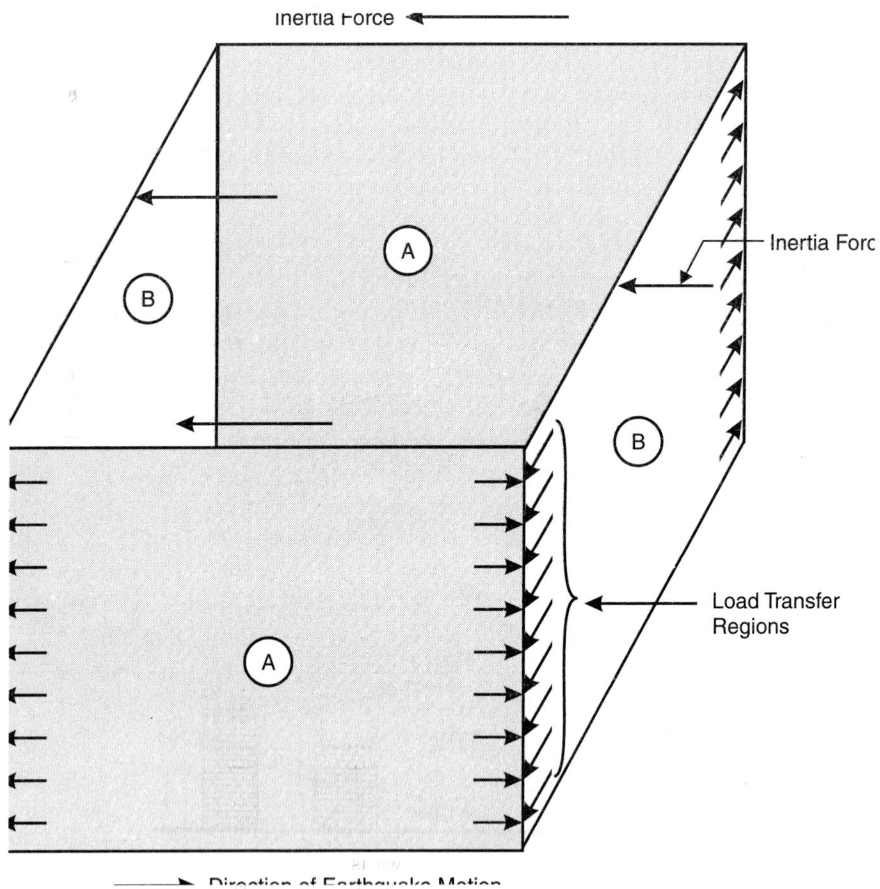

Fig. 3.6. Effect of openings in walls.

Thus the walls transfer loads to each other at their junction. Hence, the walls meeting at corners must have good interlocking. For this reason, openings near corners of the wall can be very dangerous because

(1) they will obstruct the flow of forces from one wall to another.

(2) the large openings in shear walls reduces the strength of wall against the inertia forces. (weight and area is reduced).

This type of failure results in diagonal cracks in the areas of masonry between openings and cracks at the levels of openings also. Thus, it is best to keep all opening small and away from the corners. Figure 3.7 shows the crack pattern and damage due to openings in the masonry wall.

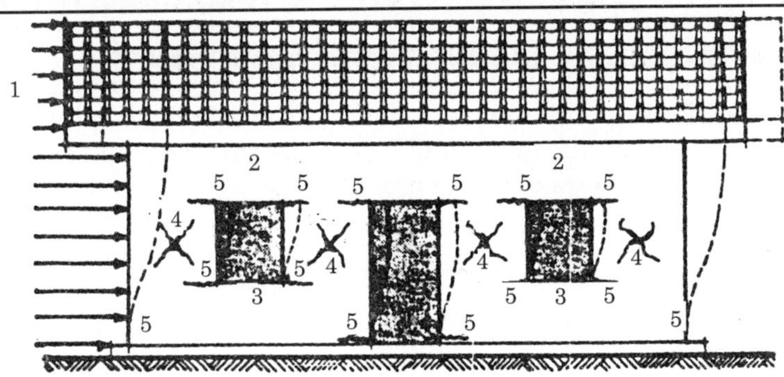

1. Earthquake force
2. Masonry
3. Sill masonry
4. Shear cracks (diagonal tension)
5. Bending cracks

Fig. 3.7. Damage due to opening.

3.6.6. Pounding

When the roofs of two adjacent buildings are at different levels then, during earthquakes, the brickwork faces hit against each other this is called as pounding. Pounding results into cracking of the walls. This failure is shown in Fig. 3.8.

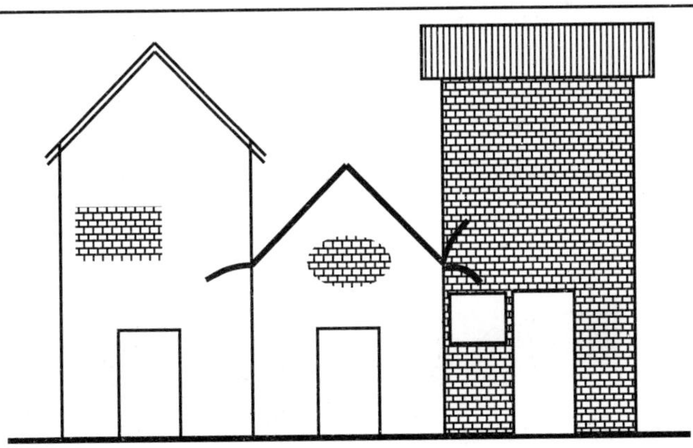

- Vertical cracks in the adjacent walls
- Vertical cracks due to different levels in the structures

Fig. 3.8. Pounding damage.

3.6.7. Non Structural Components Failure

Non-structural failure is that due to which strength and stability of the building is not affected. Such damage occurs generally under moderate earthquakes. Some non-structural, components failure are listed below :

(1) Falling of plaster from walls and ceiling.

(2) Cracking and overturning of masonry parapets, roof chimneys and balconies.

(3) Cracking and overturning of partition walls.
(4) Cracking of glass panes.
(5) Falling of loosely placed objects, overturning of cupboards etc.

3.7. TYPES OF DAMAGES OBSERVED IN TRADITIONALLY BUILT CONSTRUCTIONS DURING PAST EARTHQUAKES

The traditionally built masonry constructions have proved to be the most vulnerable to earthquake forces and have suffered maximum damage in past earthquakes. The studies of the past earthquakes (Bhuj 2001, Chamoli 1999, Jabalpur 1997, Latur 1993, Uttarkashi 1991) have shown that major loss of lives was due to collape of masonry buildings. Post earthquake damage studies help us in selecting strengtening and retrofitting measures and devising methods for improving earthquake resistant design and construction techniques. Some of the typical damages of different types of masonry constructions seen during the part earthquakes are listed below :

Brick Masonry

The buildings in fired bricks, solid concrete blocks or mortar blocks have shown following types of damage and failure during some of the past earthquake.

(1) Damage and Failure of bearing walls (Fig. 3.9)

(*a*) The failure of walls is generally due to shear and bending. Shear failure causes diagonal tension or diagonal compression. Thus, cracks are generally seen at the corners of openings and sometimes at the center of wall segments. The bending failure is observed by the tension cracks, which are vertical, at the center, ends or corners of the walls. In most of the damaged buildings. The combined (shear and bending) mode of failure is observed. Which has resulted in partial or full collapse of the walls.

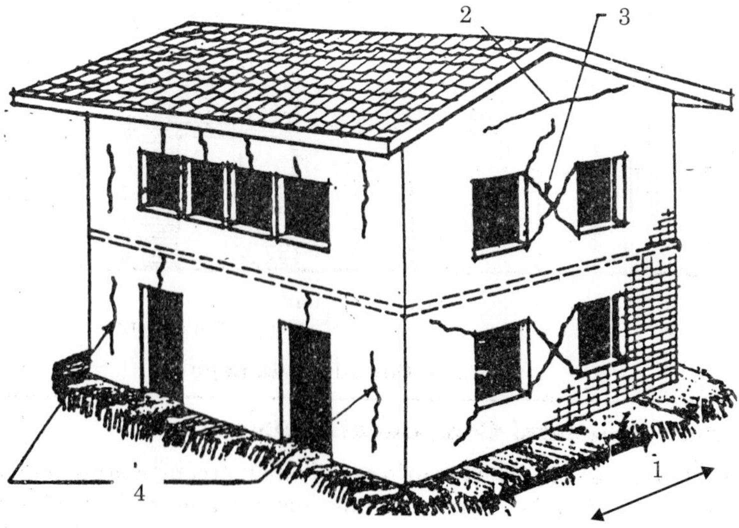

1-1 : Earthquake motion, 2-Horizontal crack in gables, 3-Diagonal cracks due to shear, 4-Cracks due to bending of wall.

Fig. 3.9. Cracking in Bearing wall due to bending and shear.

(*b*) Walls can be also damaged due to large seismic forces developed at the roof level which lead to separation of supporting walls. This type of failure is generally seen in the massive flat roofs supported by joists which are simply resting on the walls without proper connection.

(*c*) The damage in unsymmetrical buildings is more than symmetrical buildings because unsymmetry causes torsion and warping which result in large cracking in all walls.

(*d*) In the traditional constructions, arches are also very badly damaged during earthquakes.

(*e*) In the major earthquakes, wider cracks are seen which have resulted in separation of walls at corners and T junctions. This is because of lack of proper connection between walls.

(2) Failure of Ground :

(*a*) In the traditional masonry construction, the shallower depths of foundation has resulted in damage to the buildings.

(*b*) In some of the cases, where the building was constructed on loose soils, differential settlement of foundation has resulted in tilting and finally collapse of the building.

(3) Failure of Roof and Floors :

(*a*) *Dislodging of roofing material* : Loosely tied roofing material can slide due to inertia force. This failure is commonly seen in sloping roofs with slates or clay tiles as roofing material.

(*b*) Weak roof to wall connection is the cause of separation of roof from walls. In severe earthquakes, this can lead to complete roof collapse. Fig. 3.10.

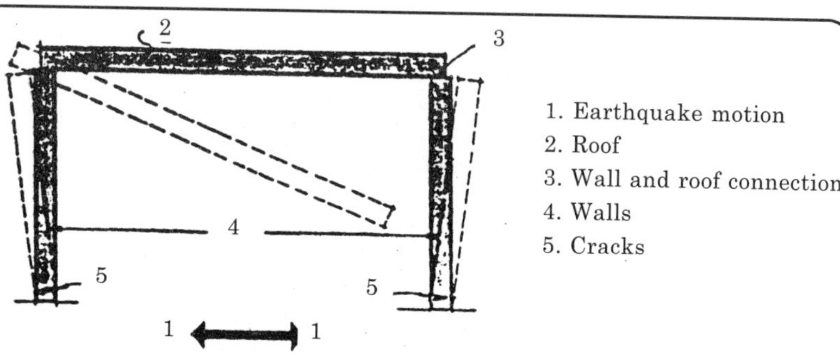

Fig. 3.10. Fall of roof because of inadequate connection between roof and wall.

3.7.1. Stone Masonry

The traditional stone masonry is used for construction since ancient times because of their availability and long lasting property. Earlier the buildings were constructed using stone and mud mortar but now people have started using cement mortar. A traditional stone house looks somewhat like shown in Fig. 3.11.

The stone masonry walls are generally very thick (700–1500 mm) and constructed with stones placed in random manner. These walls have two exterior layers called **Wythes** of large stones. The space between these vertical layers in filled with mortar and small stones (chip). The roof is also very heavy consisting of wooden planks supporting thick layer of mud. Now, due to non availability of wood some buildings use reinforced cement concrete slabs and beams with stone masonry walls. There is normally no connection between the walls and the roof.

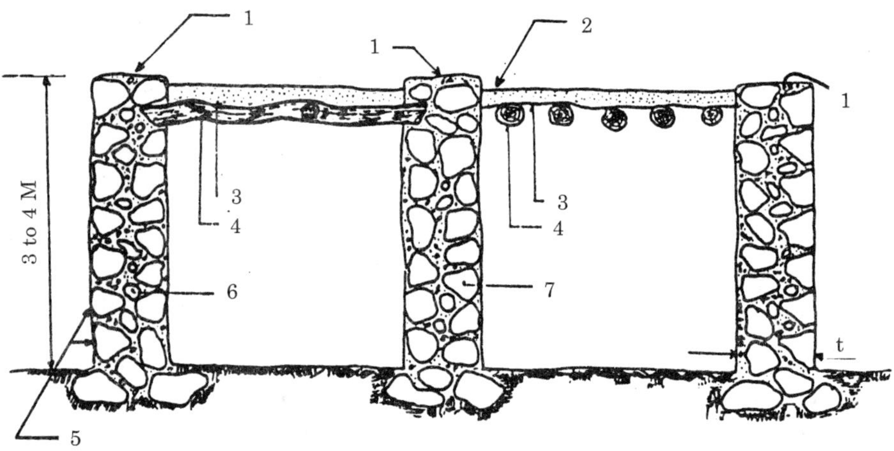

1 = Stone wall (Random rubble uncoursed masonry, 2 = Mud roof, 3 & 4 = Wooden plank, 5 = Exterior wythe, 6 = Mortar, 7 = Interior wythe, t = Thickness of wall

Fig. 3.11. Schematic cross section through a traditional stone house.

The following are the main ways in which stone masonry buildings are seen to be damaged :

(1) Separation or bulging of walls in the horizontal direction : The thick stone masonry walls get separated into two vertical wythes (layers)

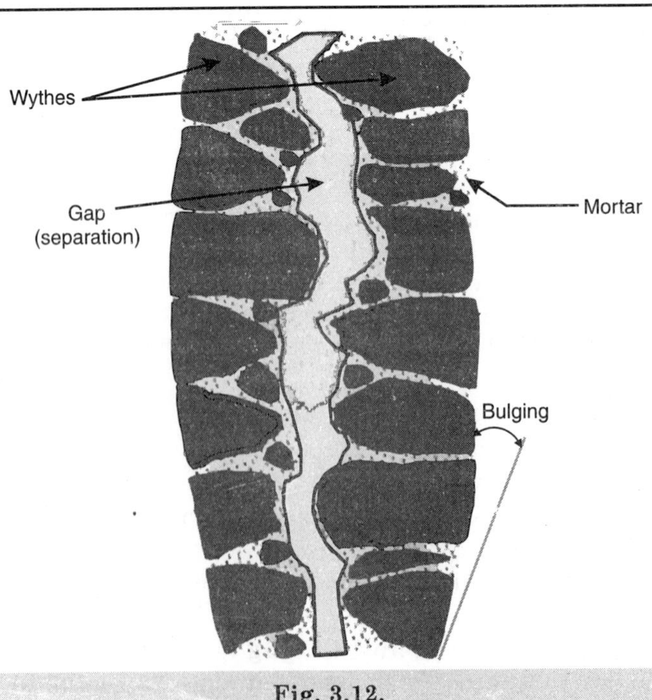

Fig. 3.12.

during earthquake shaking (Fig. 3.12). This occur mainly due toabsence of "through" stone (long stone covering inner and outer face).

(2) Separation of walls at corners and T-junctions is observed due to poorer connection between the walls.

(3) The separated wythes have collapsed under heavy weight of roof/ floors. This results in collapse of roofs causing lots of loss of lives.

Generally, the stone houses are seen completely shattered under severe earthquakes. They are reduced to heaps of stones. People get buried under them and often killed. Over 15,000 people in Bhuj, 2001 and 10000 in Latur (Maharashtra 1993) died because of collapse of stone masonry constructions. Thus, special provisions are required for making them earthquake resistant. However, looking at their dangerous collapse, it is advisable to avoid their construction in high Seismic zones.

3.7.2. Non-Engineered Reinforced Concrete Buildings

The use of reinforced concrete construction to semi urban and rural areas is very common now-a-days. The buildings are constructed using reinforced concrete columns, beams and load bearing walls. In such constructions, no consideration is given to the seismic design and connection details. Beams simply rest on columns and held in position through friction only. In this chapter focus is on low rise (upto three storeys) small buildings.

Typical damage and collapse of reinforced concrete buildings : The following types of damages are commonly observed in reinforced concrete buildings during earthquakes.

(1) Sliding of roof off the Supports : Where the beam simply rest on wall or columns, they easily slide off when earthquake intensity is more than the frictional resistance. Thus the roof many times leaves the support and fall down.

(2) Falling of Walls : The wall panels in between reinforced concrete columns were overturned during the earthquake. It happens only when they are not tightly connected with the frames.

(3) Crushing of Column ends : In many cases, the collapse of the whole framework of R.C.C. is seen. During severe shaking, the concrete (in columns) gets crushed due to heavy forces and thus, the effective sectional area gets reduced in repeated cycle of earthquake motion. The column ends become so thin that they behave like pins (Hinge) and result in collapse of whole framework.

(4) Diagonal Cracking in Columns : Due to large seismic forces the columns show diagonal cracking. The cracking results in reduction in the load carrying capacity of columns.

(5) Pulling Out of Reinforcing Bars : In many buildings, reinforcement was seen as if pulled out, cracking the concrete cover. This has happened because of inadequate development or anchor length of the bars.

(6) Foundation Sinking and Tilting : In many cases, the collapse have occured due to sinking or tilting of foundation of columns which were on loose soils. This has lead to severe cracking of the super structure and in some cases collapse also.

3.8. COMMON CAUSES OF DAMAGE

The seismic behaviour of masonry buildings during past earthquakes has revealed the strengths and weaknesses of traditionally built masonry construction. It has provided several lessons to us. Some of these are summarised below :

(1) In most of the cases, the damage and collapse has occured due to failure of connection between two walls, between walls and roof and between walls and foundation. Thus, lack of structural integrity is one of the main cause of damage leading to collapse.

(2) One of the main cause of damage is the absence of bond between perpendicular walls causing separation from each other at the junction. It results in loss of box action and making the walls weak in resisting seismic forces.

(3) The large openings in masonry walls has also resulted into severe damage. This damage is commonly seen as vertical cracks over the openings and diagonal cracks in the masonry between openings.

(4) Roofs, that are not properly connected to the walls, can separate from the walls. This may cause severe loss of lives (due to heavier roofs)

(5) Out of plane failure is more commonly observed than in plane failure.

(6) In general, buildings with irregular plans experienced more damage than rectangular buildings.

(7) Non-structural elements must be given proper design considerations for lateral forces.

(8) Thick stone masonry walls have bulged out/separated into vertical columns because of absence of "through" or bond stone.

(9) Use of lighter and regular construction is the best way to reduce damage due to earthquakes.

TERMS TO REMEMBER

1. **Masonry construction :** The construction comprising of bricks, stone, concrete blocks or clay (earth) alongwith cement sand or mud mortar (used for binding them).
2. **Shear wall :** The wall which is parallel to the ground motion and subjected to shear and bending in its plane. Shear walls are good in carrying horizontal forces.
3. **Flexural wall :** The walls which are transverse to the direction of motion and are subjected to bending in a direction perpendicular to its plane (overturning).

4. **Diaphragm :** It is a horizontal system like roof which transmits lateral forces to the vertical resisting element such as walls and columns.
5. **Out of plane failure :** Overturning of flexural wall which is the result of out of plane bending is called as out of plane failure.
6. **In plane failure :** The failure of shear wall as a result large shear and bending forces in plane of the wall.
7. **Box action :** The walls, roof and foundation of a building should be very well connected so that they act as a single unit. Under the action of horizontal forces. This is called as **Box action**.
8. **Pounding :** The hitting or hammering action of two adjacent buildings of different heights during earthquake.
9. **Whythes :** The thick stone walls have two vertical layers of large stones and the space between them is filled with mortar and small stones. The two layers of stones are called as Wythes.
10. **Through stone :** It is a long stone which cover the entire thickness of the stone wall (inner and outer face).

EXERCISES

A. Objective Type Questions :

1. Which of the following are the earthquake effects :
   (*a*) Tsunamis (*b*) Ground motion
   (*c*) Fire (*d*) All of the above
2. Which of the following is not traditional construction.
   (*a*) Steel frame buildings (*b*) Brick masonry
   (*c*) Stone masonry (*d*) Earthen buildings
3. Advantages of masonry construction are
   (*a*) Cheap and easy
   (*b*) Thermal comfort and sound insulation
   (*c*) Heavy structure (*d*) (*a*) and (*b*)
   (*e*) (*a*), (*b*) and (*c*)
4. Masonry construction has
   (*a*) High tensile strength and low shear strength
   (*b*) Low tensile strength and low shear strength
   (*c*) High tensile strength and high shear strength
   (*d*) Low tensile strength and high shear strength
5. The tendency of a wall to overturn is increased with
   (*a*) More length (*b*) More thickness
   (*c*) No effect (*d*) (*a*) and (*b*)

6. Shear walls are subjected to
   (*a*) Out of plane and bending (*b*) In plane bending
   (*c*) Both (*a*) and (*b*) (*d*) No bending
7. Box action provides greater resistance to
   (*a*) Horizontal loads (*b*) Vertical loads
   (*c*) Line loads (*d*) All of the above
8. In a building "Diaphragm" term is used for
   (*a*) Walls (*b*) Beams
   (*c*) Roof (*d*) Columns
9. Long walls are easier to
   (*a*) Slide (*b*) Crack
   (*c*) Overturn (*d*) All of the above
10. Reinforced concrete slab acts as
   (*a*) Rigid diaphragm (*b*) Flexible diaphragm
   (*c*) Brittle diaphragm (*d*) None of the above
11. Bending failure in masonry is observed by
   (*a*) Vertical cracks (*b*) Diagonal cracks
   (*c*) Horizontal crack (*d*) All of the above
12. Shear failure in masonry is observed by
   (*a*) Vertical cracks (*b*) Diagonal cracks
   (*c*) Horizontal crack (*d*) All of the above
13. Unsymmetrical buildings show more damage during earthquakes because of
   (*a*) Bending (*b*) Shear
   (*c*) Torsion and warping (*d*) None of the above
14. Stone walls, have two layers, called as Wythes generally fail, during earthquakes, because of
   (*a*) Bending of wall
   (*b*) Low shear strength
   (*c*) Bulging or separation of walls into two layers
   (*d*) All of the above
15. Separation of walls at corners and T junctions is observed due to
   (*a*) Less mortar
   (*b*) Low shear strength of masonry
   (*c*) Low tensile strength of masonry
   (*d*) Poor connection between the walls
16. Pulling out of reinforcing bars in RCC construction occur due to
   (*a*) Less development length
   (*b*) Poor quality of steel

(*c*) Poor quality of concrete

(*d*) None of the above

17. Loose soils may result in

(*a*) Differential settlement of building

(*b*) Sliding off of the super structure

(*c*) Large inertia forces

(*d*) None of the above

18. Bulging of walls in stone masonry can be prevented by

(*a*) Good mortar

(*b*) Heavy stones

(*c*) By providing "through" stone

(*d*) All of the above

B. Fill in the Blanks :

1. Masonry has very ________ tensile strength and ________ shear strength.
2. In masonry construction, stress concentration occurs at ________ of windows and doors.
3. The walls which are transverse to the direction of motion are called as ________.
4. The walls which are parallel to the direction of motion are called as ________.
5. The flexural walls are more prone to ________ failure.
6. Tall and slender walls are more prone to ________ during earthquake.
7. A wall with ________ length to width ratio develops horizontal cracks due to tension caused by bending.
8. The box action provides ________ resistance to the horizontal loads.
9. ________ failure occurs if the walls are not tied together properly.
10. ________ failure is common in the roofs consisting of timber joists with brick tiles.
11. Openings in walls should be made ________ from the corners.
12. Bending failure can be observed in the form of ________ tension cracks.
13. Earthquake damage is seen ________ in unsymmetrical building than symmetrical building.
14. For better seismic resistance, the walls should have ________ interlocking.
15. Loose soil result in ________ of foundation and tilting leads to collapse of building.
16. Earthquake shaking leads to ________ of thick walls in stone masonry construction.
17. Bulging of walls can be prevented by providing ________ stone in the walls.

18. _________ is an excellent material for earthquake resistance or rehabilitation.
19. _________ weight earthen construction cause great loss of lives and damage during earthquakes.
20. Dry straw is used in mud mortar to control _________.
21. _________ and _________ construction is the best way to reduce damage due to earthquakes.
22. The hitting action of two adjacent buildings of different heights during earthquakes is called as _________.
23. The simplest way to make masonry buildings earthquake resistant is by tying or bonding the walls and roof properly so that they act as a _________.
24. Shear failure is observed as _________ cracks.
25. Openings in walls _________ the strength of walls against inertia forces.

C. State True/False :

1. Soil liquefaction general occurs in unsaturated soils.
2. Masonry construction provide good insulation against heat.
3. Masonry has low shear strength and good tensile strength.
4. Flexural walls are prone to overturning during earthquakes.
5. In plane failure is commonly observed in long shear walls.
6. Vertical cracks in the corners of walls and at T junctions shows out of plane failure.
7. Diagonal tension cracks are seen in short piers between openings in plane failure.
8. Diaphragm failure is the failure of foundation.
9. Box action provides greater resistance to horizontal loads.
10. Openings should be placed near the corners to prevent failure.
11. Pounding results into cracking of the walls.
12. Symmetrical buildings are more prone to damage during earthquake.
13. Stone masonry construction is not advisable in high seismic zones.
14. Out of plane failure is more commonly observed than in plane failure during earthquakes.
15. Wythes are the "through" stone used in stone masonry construction.
16. Reinforced concrete slab act as rigid diaphragm.
17. Masonry shows ductile behaviour in tension.
18. Non-structural failure is that due to which strength and stability of the building is not affected.
19. Diagonal bracing reduces the damage due to earthquake in a wooden building.
20. Non-engineered buildings are not made of reinforced cement concrete.

21. The horizontal force resisting elements (roof slab and floors) are called as diaphragms.
22. Masonry buildings are less prone to damage than reinforced concrete buildings.
23. Long and slender walls are more damaged during earthquake.
24. Partition walls are the non-structural component of a building.
25. Traditionally built construction is more rigid.

ANSWERS

A. Objective Type Questions :

1. (*d*)	2. (*a*)	3. (*d*)	4. (*b*)	5. (*a*)
6. (*b*)	7. (*a*)	8. (*c*)	9. (*c*)	10. (*a*)
11. (*a*)	12. (*b*)	13. (*c*)	14. (*c*)	15. (*d*)
16. (*a*)	17. (*a*)	18. (*c*)		

B. Fill in the Blanks :

1. low, low	2. corners	3. flexural	4. shear	5. bending
6. overturning		7. small	8. good	9. Connection
10. diaphragm		11. away	12. Vertical	13. more
14. good		15. differential settlement		16. bulging
17. Through	18. Timber	19. Heavy	20. microcracking	
21. Lighter, symmetrical		22. Pounding	23. Box	
24. diagonal tension		25. reduces		

C. State True/False :

1. False	2. True	3. False	4. True	5. True
6. True	7. True	8. False	9. True	10. False
11. True	12. False	13. True	14. True	15. False
16. True	17. False	18. True	19. True	20. False
21. True	22. False	23. True	24. True	25. False

SHORT ANSWER TYPE QUESTIONS

1. What are the materials of traditional construction used in India ?
2. Define the terms shear walls and flexural walls.
3. What is box action ?
4. How can connection failure be prevented in a building ?
5. Define diaphragm failure ?
6. Where should the opening be placed in wall for good earthquake resistant construction ?
7. What is pounding ?

8. Name modes of failure observed in flexural walls and shear wall ?
9. What do you understand by the term 'Wythes' ?
10. Why does a building constructed on loose soil collapse during earthquake ?
11. How can bulging of stone masonry walls be prevented ?
12. Why is the use of stone masonry not advisable in high seismic zones ?

LONG ANSWER TYPE QUESTIONS

1. What are the effects of earthquake ?
2. (*a*) Discuss some of the traditionally built constructions in India.
   (*b*) Discuss the merits and demerits of masonry constructions
3. Explain the terms flexural wall and shear wall with the help of a diagram.
4. Enumerate the various failure mechanism of a masonry building.
5. What is the difference between out of plane failure and in-plane failure of a masonry building ?
6. What do you mean by rigid and flexible diaphragms ? What causes the failure of diaphragm ?
7. How does the openings in walls affect their performance during earthquake motion ?
8. What is box action ? How does it help in transfer of forces from one wall to another ?
9. Explain connection failure in a masonry building. How can we prevent it ?
10. (*a*) Define Pounding of buildings.
    (*b*) What are the various types of non-structural component failure ?
11. Discuss different types of damages observed in Brick masonry building.
12. Discuss various ways in which stone masonry buildings are damaged during earthquakes.
13. (*a*) What is meant by non-engineered reinforced concrete construction ?
    (*b*) What are the various types of damages commonly observed in non-engineered R.C.C. buildings ?
14. Explain various causes of damage and failure of masonry buildings.

❑❑❑

4

CHAPTER

Introduction to Seismic Design Codes : IS 4326:2013 and IS 13828:1993

4.1. INTRODUCTION

Masonry buildings are the most common type of traditional construction used for housing purpose all around the world. These are mostly seen in rural, urban and hilly regions. This type of construction has very low seismic resistance. The recent earthquakes have shown that the collapse of these construction is the main cause of destruction. Hence, it is necessary to increase the seismic resistance of these constructions.

In most cases, the resistance can be improved by following simple, inexpensive principles of good building construction. But already constructed in use buildings are of concern. Thus, retrofitting of such building is also essential for improving their seismic resistance. Therefore, study of all these features becomes a must for every civil engineer. This chapter explains various provisions given in the Indian Standard Codes for improving the Seismic resistance of various types of traditionally built constructions like brick masonry. The chapter also focusses on the General principles to be followed for constructing earthquake resistant buildings.

4.2. EARTHQUAKE RESISTANT CONSTRUCTION

An earthquake resistant construction should have following properties for better seismic performance :

(1) Good Structural Configuration

(2) Lateral Strength

(3) Ductility

(4) Deformability

(5) Damageability.

(1) Good Structural Configuration : The size, shape and structural system of building should be such that inertia forces are transferred to the ground safely. An important property of good structural configuration is its integral action. Integral action means the structure acts as one unit and is able to resist the earthquake forces in a better and safer way.

(2) Lateral Strengh : The earthquake result in large inertia forces. Thus, a good earthquake resistant building should have enough lateral strength, so that, it can resist the inertia forces without collapse.

(3) Ductility : Ductility is the property of a material which enable it to undergo large elongation before breaking. The materials generally used for construction are masonry concrete and steel. Masonry and concrete are brittle while steel is ductile. A good earthquake resistance building should have enough ductility. This can be done by addition of ductile material such as wood in earthen construction and steel bars in masonry and concrete construction.

(4) Deformability : Deformability refers to the ability of a structure to undergo large deformations without collapse. While ductility is the inherent property of a material, deformability pertains to the structure. The deformability of a structure can be increased by making regular, well proportional structure which is tied properly. Good deformability of a structure is also necessarity to make it earthquake resistant . For example– Even when ductile material in present in the building components, such as beams and walls, the building may fail if the components are not well proportioned and tied together properly.

(5) Damageability : Damageability refers to the ability of a structure to undergo substantial damage without partial or total collapse. This will result in sufficient warning to the people before collapse, thus resulting in less loss of lives. Good damageability can be achieved by providing several supports to important structural components (such as beams) and avoiding central columns or walls supporting large portions of building.

4.3. TRADITIONALLY BUILT MASONRY CONSTRUCTIONS

Masonry constructions have low earthquake resistance. Various types of damage commonly seen in masonry buildings and their causes are studied in the last chapter. On the basis of them, some of the factors responsible for low seismic resistance of masonry buildings are as follows :

1. Lack of integral action.
2. Lack of strong and ductile connections between walls, walls and roof, and walls and foundation.
3. Less strength for out of plane forces.

4. Low tensile and shear strength of masonry.
5. High in-plane stiffness of wall.
6. Low ductility and deformability.
7. Heavy mass of the structure.

All these weaknesses can be overcome by following simple principles and methods given in various codes for the earthquake resistant design and constructions of masonry buildings.

4.4. SEISMIC DESIGN CODES

Bureau of Indian standard has published various Seismic Design Codes which gives guidelines for design and construction of earthquake resistance structures. These codes take into consideration the effect of earthquake mainly forces and deformations. They also take into account the local seismology, building type, materials and methods of construction used in a particular region or country. The first Indian seismic code was published in 1962 as IS : 1893. Today, Bureau of Indian Standads has published following seismic codes, covering different types of constructions/structures :

(1) IS : 1893 (Part 1 to 5) 2016 Indian Standard for Earthquake Resistant Design of Structures (6th revision)

(2) IS : 4326, 2013 Indian Standard Code of Practice for Earthquake Resistane Design and Construction of Buildings (3rd revision).

(3) IS : 13827, 1993 Indian Standard Guidelines for Improving Earthquake Resistance of Earthen Buildings.

(4) IS 13828, 1993 Indian Standard Guidelines for Improving Earthquake Resistance of Low Strength Masonry Buildings.

(5) IS : 13920, 2016 Indian Standard Code of Practice for Ductile Detailing of Reinforced Concrete Structures Subjected to Seismic Forces.

(6) IS : 13935, 2005 Indian Standard Guidelines for Repair and Seismic Strengthening of Buildings.

These codes do not ensure that structures suffer no damage during earthquakes of all magnitudes. But they ensure that structures are able to with stand earthquake shaking of moderate intensities without much damage and heavy intensities without total collapse.

This chapter focusses on the traditionally built constructions and hence covers the codes related with them only. These codes define methods and techniques which help the engineers in planning, designing, detailing and constructing these structures. The codes related with traditional masonry constructions are :

1. IS : 4326 : 2013
2. IS : 13828 : 1993
3. IS : 13827 : 1993

SECTION-1

4.5. INTRODUCTION TO IS 4326 : 2013

The code covers general principles for earthquake resistant building construction. This code also deals with the selection of materials and special features of design and construction for the following types of buildings.

- Masonry construction using rectangular masonry units.
- Timber constructions.
- Buildings with prefabricated reinforced concrete roofing/flooring elements.

Various codal provision of IS4326, 2013 are explained in the following articles.

4.6. GENERAL PRINCIPLES FOR EARTHQUAKE RESISTANT BUILDINGS

The earthquake resistance of buildings can be improved by following simple principles of design and good building construction practices. These principles are given in the code and are explained below :

1. Lightness : The earthquake force depends on mass of the structure. Heavier structure means large inertia force and collapse of these structures results in heavier damage and loss of lives. **Thus, a building should be as light as possible, especially the roofs and upper storeys.**

2. Building Configuration : The behaviour of a building during earthquake depends on its shape, size and geometry. A good building configuration can result in less damage during earthquakes. The various components of building configuration are explained below :

(*a*) Symmetry : The building as a whole or its various blocks should be kept symmetrical about both the axes. The asymmetrical buildings are subjected to twist or torsion during earthquakes. This twist make different portions at the same floor level to move horizontally by different amounts. This causes more damage. This damage can be minimized by planning symmetrical buildings as shown in Fig. 4.1.

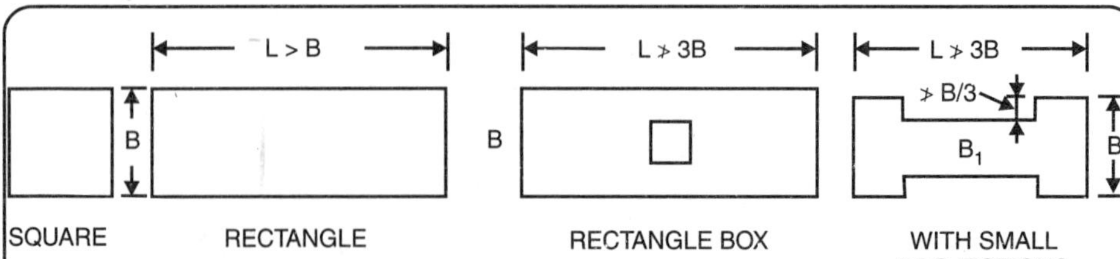

Fig. 4.1. Symmetrical desirable plans.

If asymmetry is not avoidable than the additional forces due to torsion should be taken while designing the structures.

Before Earthquake

After Earthquake

Damaged Historical Monument, Nepal Earthquake, 2015

(b) Simplicity and Regularity : The building should have a simple rectangular plan. It is seen that simple shapes behave better during earthquake than complex shapes like L, T, E, H, U and T etc. shown in Fig. 4.2. It is seen that during earthquakes the buildings with **re-entrant corners** have suffered great damage. These type of buildings can be broken into rectangular blocks which are separated properly. **Thus, separation of a large building into smaller blocks can lead to symmetry and regularity** as shown in Fig. 4.3.

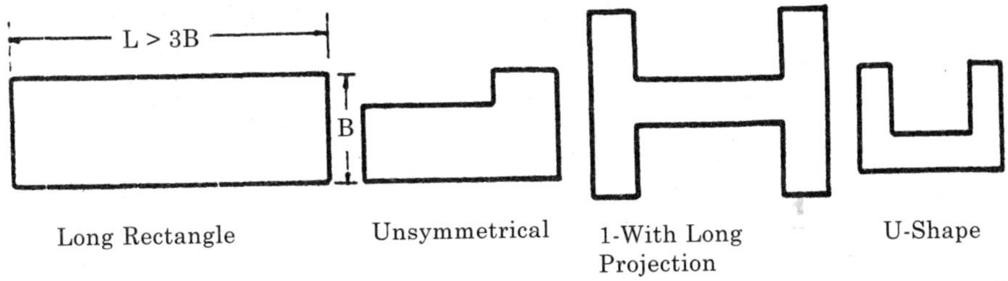

Fig. 4.2. Long or unsymmetrical undesirable plans.

For preventing pounding or hammering (Art. 3.6.6) between blocks, a separation of 3 to 4 cm throughout the height above plinth level is required. This separation section is just like an expansion joint or it may be filled with a weak material which can easily crush during earthquake shaking.

(c) Simple Building without much Projections and Suspended Parts behave well during earthquake : Long cornices, vertical or horizontal projections, facia stones etc. should be avoided and are dangerous during earthquakes. If these parts cannot be avoided, they should be reinforced properly and tied firmly to the main structure. Ceiling plaster should not be done and if done, it should be as thin as possible.

(d) Size of the Building : In tall buildings, the horizontal movement of the floors during ground shaking is large. In short, but very long buildings the damaging effects of earthquake are more. **Buildings with one of their dimension much larger or smaller than the other two do not perform well during earthquakes.** Thus the buildings length should not be more than three times its width. If longer lengths are needed, two separate blocks with separation should be provided (Fig. 4.3).

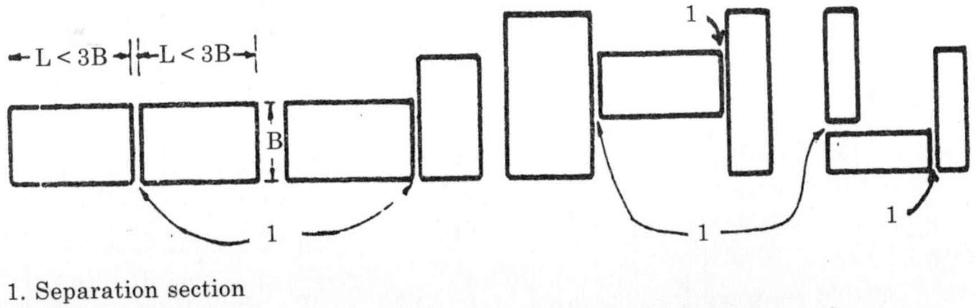

Fig. 4.3. Use of separation section for improving plans.

(*e*) Enclosed Area : A small building with properly interconnected walls acts like a rigid box and more earthquake resistant. Therefore it is advisable to have separate small rooms than one long room.

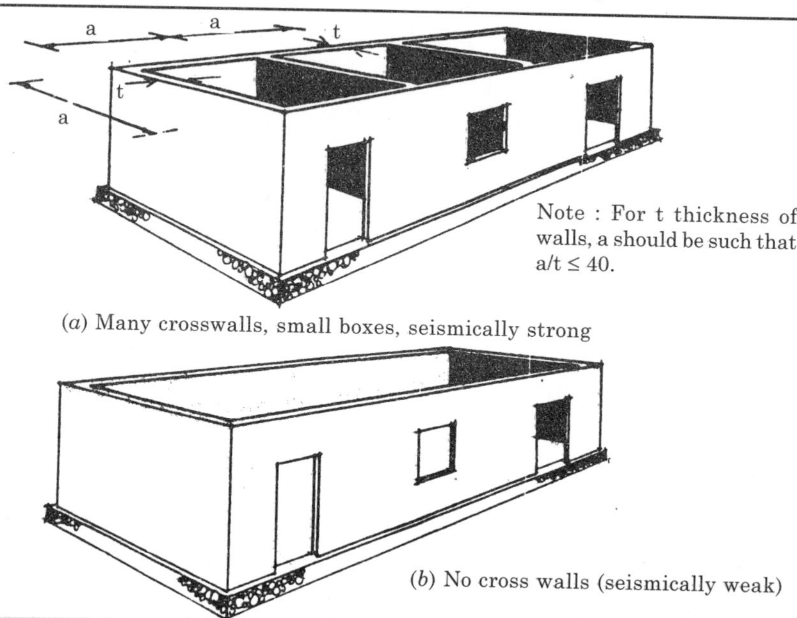

Fig. 4.4.

3. Strength in Various Directions : The structure should have adequate strength along both the axes. The design should also be safe and take into account the reversible nature of earthquake motion.

4. Stability of Slope : Hill side slopes are liable to slide during an earthquake. Hence buildings should not be constructed on them only stable slopes should be chosen. Similarly buildings with unequal members will also twist under ground movement and may result in damage and collapse.

5. Foundation : The buildings should not be constructed on loose soils. These soils will compact and subsidise or liquefy and result in unequal settlement of building and damage it.

Although such soil can be compacted properly for small buildings. But for large buildings this operation is costly. For large buildings, rigid raft foundation or piles taken to firm stratum can be used.

6. Ductility : Ductility is the most desirable property for good earthquake performance. The brittle masonry can be strengthened by providing steel reinforcing bars at critical sections which will also improve ductility of the structure.

7. Fire Resistance : It is very common to see fire after earthquake. It may be because of electrical short circuits, leaking of gas pipes/cylinders kerosene lamps and kitchen fires. The fire hazard is sometimes more serious than the earthquake damage. Hence, the buildings should be made fire resistant according to Indian standards (IS 1641, 1642, 1643, 1644 and 1646).

4.7. SPECIAL CONSTRUCTION FEATURES (CLAUSE 5)

1. Separation of adjoining structures: It is required for structures having unequal storey heights and different dynamic characteristics. This is to avoid collision during an earthquake.

2. Roof and Floors : Flat roof or floor should not be made of tiles or bricks supported on steel, timber or reinforced concrete joists. There bricks or tiles can be loosened and may fall during earthquake.

For pitched roofs, corrugated iron or asbestos sheets should be used in place of tiles or other loose roofing units. All roofing material should be tied properly to the supporting members. Heavy roofing material should be avoided.

3. Stair Cases :

- Large stair hall should be separated from the rest of the building.
- The interconnection of the stairs with the adjacent floors shold be provided by sliding joints at the stairs to prevent the bracing effect on the floors.

4.8. CATEGORIES OF BUILDINGS

As per IS : 4326, 2012. For the purpose of specifying the earthquake resistant features in masonry and wooden buildings, the buildings, have been categorized in four categories, B to E based on the seismic zone and the importance of building I, where

I = Importance factor applicable to the buildings given in Table 4.1.

TABLE 4.1. Importance Factors.

Sl. No. (1)	Structure (2)	Importance Factor (3)
(*i*)	Important service and community buildings, such as hospitals; schools; monumental structures; emergency buildings like telephone exchange, television stations, radio stations, railway stations, fire station buildings; large community halls like cinemas, assembly halls and subway stations, power stations	1.5
(*ii*)	All other buildings	1.0

Notes :
1. The design engineer may choose values of importance factor I greater than those mentioned above.
2. Buildings not covered in Sl. No. (*i*) and (*ii*) above may be designed for higher value of I, depending on economy, strategy considerations like multi-storey buildings have several residential units.
3. This does not apply to temporary structures like excavations, scaffolding etc. of short duration.

TABLE 4.2. Building Categories for Earthquake Resistant Features.

Importance Factor	Seismic Zone			
	II	III	IV	V
1.0	B	C	D	E
1.5	C	D	E	E

Note. Category A is now abandoned as zone I does not exist any more.

4.9. CODAL PROVISIONS OF IS : 4326 : 2013

Various codal provisions of IS : 4326 : 2013 regarding material selection, design and construction of earthquake resistant buildings are as follows :

1. Masonry Unit :

(*i*) For earthquake resistant buildings, well burnt bricks (IS 1077) and solid concrete blocks (IS 2185) having crushing strength not less than 3.5 MPa should be used.

(*ii*) Squared stone masonry, stone block masonry as per IS : 1597 (Part- 2) : 1992 may also be used.

2. Mortars :

(*i*) Since tensile and shear strength are important for earthquake resistance, use of mud or very weak mortar is not suitable. A mortar mix of cement : Sand (1 : 6) by volume at least should be used. Table 4.2 gives the list of recommended mortar mixes for various categories of buildings.

(*ii*) Where steel reinforcing bars are provided in masonry, the bars shall have proper cover in cement : sand mortar (not less than 1 : 3) and the minimum clear cover is 10 mm. In cement concrete of grade M20, minimum clear cover is 15 mm or bar diameter whichever is more.

3. Walls :

(*i*) The thickness of load bearing walls should not be less than 190 mm and wall length should not be greater than 15 m.

(*ii*) The masonry bearing walls can be built up to a maximum of four storeys.

(*iii*) The masonry bearing walls should be properly reinforced.

(*iv*) The bearing walls should be straight and symmetrical in plan in both directions.

(*v*) The wall panels should be checked for strength in bending and shearing.

4. Masonry Bond :

(*i*) For achieving full strength of masonry the usual bonds specified for masonry should be followed.

(*ii*) The vertical joints should be broken properly from course to course.

(*iii*) For having full bond between perpendicular walls the joint should be made sloping (stepped) by making the corners first to a height of 600 mm and then building the wall in between them. Otherwise the toothed joint should be made in both the walls alternatively in lifts of about 450 mm as shown in Fig. 4.5.

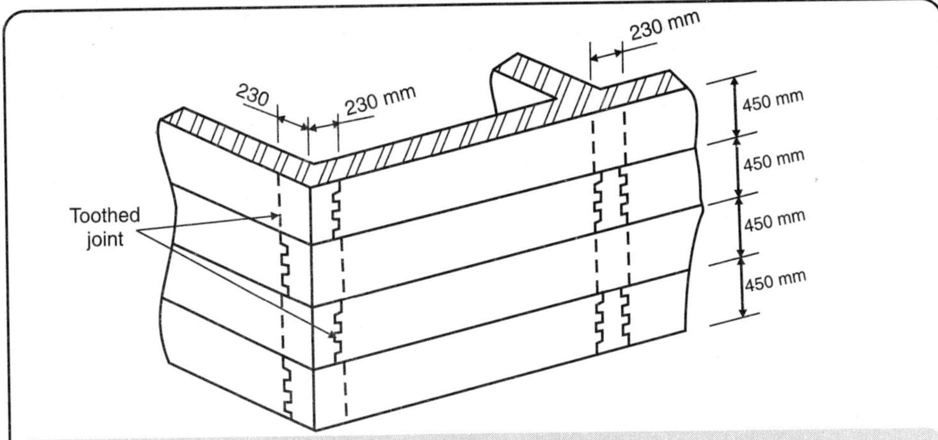

Fig. 4.5. Toothed joints in walls at corners and T junctions.

5. Openings in Walls : The following are the guidelines on the size and position of openings in walls.

(*i*) As openings in walls reduce their lateral load resistance so they should be small in size and centrally located. The guidelines on the size and position of openings are given in Table 4 of IS 4326:2013.

(*ii*) Openings in any storey should have their top at the same level so that a continuous band could be provided over them.

(*iii*) If openings are not provided according to the codal provisions, they should be strengthened by reinforced concrete lining with high strength deformed bars of 8 mm diameter as shown in Fig. 4.6.

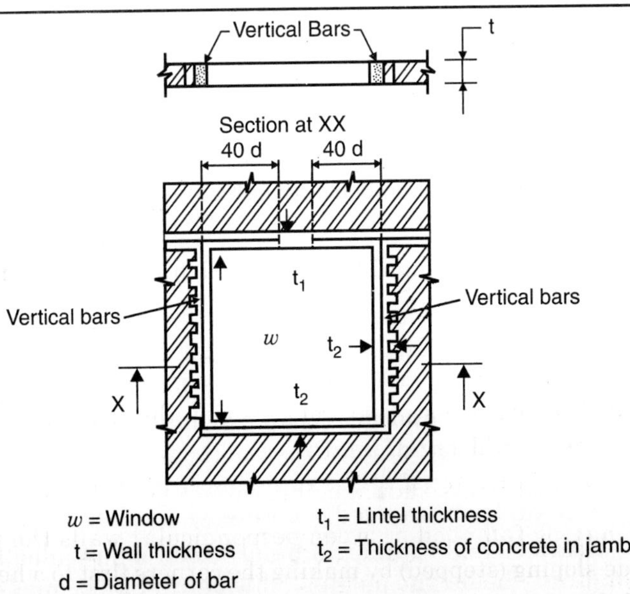

Fig. 4.6. Strengthening Masonry Around Openings.

(*iv*) If a window or ventilator is to be projected out, the projection should be in reinforced masonry or concrete and tied together properly.

(*v*) Use of arches over the openings should be avoided. If arches are provided then steel ties should be provided.

4.10. SEISMIC STRENGTHENING ARRANGEMENTS (CLAUSE 8.4)

All masonry buildings should be strengthened in horizontal as well as vertical direction for improving the earthquake resistance. The strengthening arrangements are disucssed in clause 8.4 of IS : 4326. These are very important and are explained below.

(1) Horizontal reinfrocement

(2) Vertical reinforcement.

4.10.1. Horizontal Reinforcement

The horizontal reinforcing of walls is required for imparting them horizontal bending strength against inertia force. It also helps in tying the walls together. In the exterior walls horizontal reinforcement helps in preventing shrinkage and temperature cracks. The following arrangements of horizontal reinforcement are necessary for earthquake resistant buildings.

(*a*) Horizontal Bands or Ring Beams. The most important seismic. strengthening arrangement for buildings is through reinforced concrete bands. **A band is reinforced concrete or reinforced brick runner provided in the walls to tie them together and to impart horizontal bending strength to them.**

These bands are provided continuous through all the load brearing walls at plinth, lintel, roof caves level and also at top of gables. The various bands are explained below :

(*i*) **Lintel Band :** This is the most important band and provided at lintel level in all load bearing and all cross walls. **It must be provided in all storeys of building.** Lintel band ties all the walls together and resist out of plane bending of the walls. The lintel band also improve the stability of walls during severe earthquake. The section and reinforcement details of lintel band are given in Table 4.6.

(*ii*) **Plinth Band :** This band should be provided in those cases where the soil is soft or uneven as in hily areas. This band helps in reducing the differential settlement of building. This band will also serve as damp proof course. This band is provided at plinth level of walls on top of the foundation wall. Its section is same as lintel band.

(*iii*) **Roof Band :** Roof band is a band provided immediately below the roof or floors. This band improves the resistance against diaphragm failure thus it is a must for flexible diaphragms. This band need not be provided under reinforced concrete or reinforced brick roof.

(*iv*) **Gable Band :** Gable band is a band which is provided at the top of gable masonry below the purlins. This band shall be made continuous with the roof band at the eaves level. This band resists out of plane failure of gable walls.

Section and Reinforcement of Band

The band should be made of reinforced concrete of grade not less than M15 or reinforced brick work in cement mortar not leaner than 1 : 3.

(1) The band should be of the full width of wall.

(2) The depth of band should not be less than 75 mm and reinforced with steel, as given in clause 8.4.5 of IS 4326, Table 6.

(3) In case of reinforced brickwork, the thickness of joints containing steel bars shall be increased so as to have a minimum mortar cover of 10 mm around the bar.

(4) In bands of reinforced brickwork, the area of steel provided should be equal to that for reinforced concrete bands.

(5) The continuity of bands at corners and function of walls is must for integral action. The details of bands at corners and junctions are shown in Fig. 4.7.

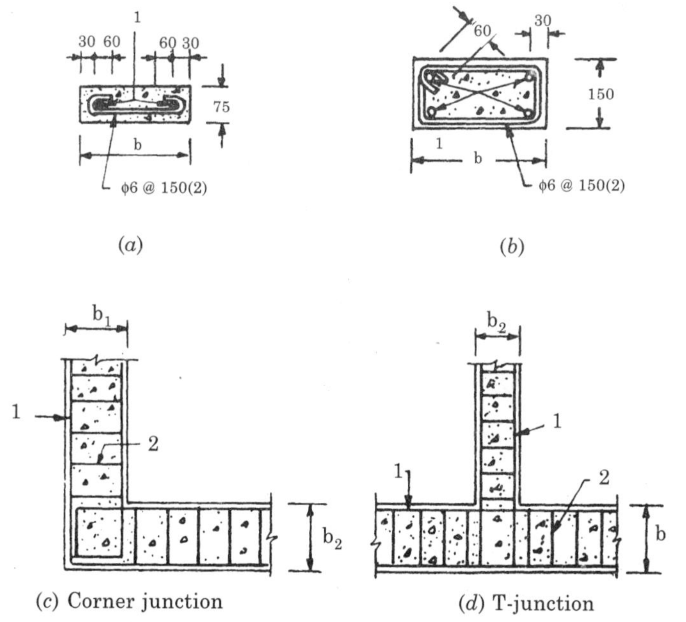

1. Longitudinal bars, 2. Lateral ties, b_1, b_2 = Wall thickness
All dimensions in millimetres

Fig. 4.7. Reinforcement details in band at corners and junctions.

(*b*) Dowel Bars : Dowel bars are provided in category D & E of buildings. These are provided to improve the BOX ACTION of walls. Steel dowel bars are provided at corners and T junctions of walls at the sill level of windows to a length of 900 mm. These dowel bars are in the form of U stirrups of 8 mm diameter. These bars must be laid in 1 : 3 cement sand mortar with a minimum cover of 10 mm (Fig. 4.8)

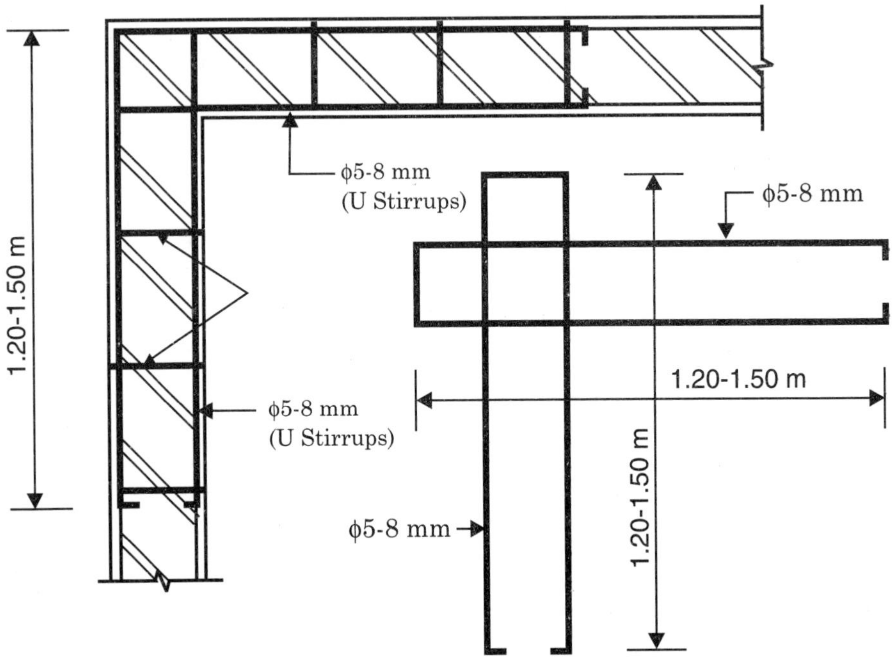

Fig. 4.8.

4.10.2. Vertical Reinforcement (Clause 8.4.8)

Vertical reinforcement is also provided in walls to improve the seismic resistance of buildings. Various points to be considered for vertical reinforcement are as follows

(1) Vertical steel at corners and junctions of walls which are up to 340 mm $\left(1\frac{1}{2}\text{ brick}\right)$ thick shall be provided as per Table 7 of IS 4326.

(2) For walls thicker than 340 mm area of bars can be increased proportionately. The amount of vertical steel depends upon number of storeys and category of building. No vertical steel is provided in category A building.

(3) The vertical reinforcement should be property embedded in the plinth masonry and roof slab or roof band.

(4) The vertical reinforcement should pass through the lintel bands and floor level bands in all storeys.

(5) Vertical reinfrocement for window and door openings should start from foundation of floor and upto lintel band.

(6) The typical details of providing vertical steel in brick masonry buildings are shown in Fig. 12 of IS 4326.

TABLE 4.3. Strengthening Arrangements Recommended for Masonry Buildings (Rectangular Masonry Units) (Clause 8.4.1)

Building category (1)	Number of storeys (2)	Strengthening to be Provided in all Storeys (3)
B	(*i*) 1 to 3	a, b, c, f, g
	(*ii*) 4	a, b, c, d, f, g
C	(*i*) 1 and 2	a, b, c, f, g
	(*ii*) 3 and 4	a to g
D	(*i*) 1 and 2	a to g
	(*ii*) 3 and 4	a to h
E	1 to 3*	a to h

where a = Masonry mortar

b = Lintel band

c = Roof band and gable band where necessary

d = Vertical steel at corners and junctions of walls

e = Vertical steel at jambs of openings

f = Bracing in plan at tie level of roofs

g = Plinth band where necessary

h = Dowel bars

*4th storey not allowed in category E.

Note : In case of four storey buildings of category B, the requirements of vertical steel may be checked coefficient equal to four times the one given in (a) 3.4.2.3 of IS 1893 : 2016. (This is because the brittle behaviour of masonry in the absence of a vertical steel results in much higher effective seismic force than that envisaged in the seismic coefficient, provided in the code). If this analysis shows that vertical steel is not required the designer may take the decision accordingly.

The total strengthening arrangement of providing reinforcing steel (horizontal and vertical bands) in masonry buildings is listed in Table 4.3 and in Figs. 4.9, 4.10.

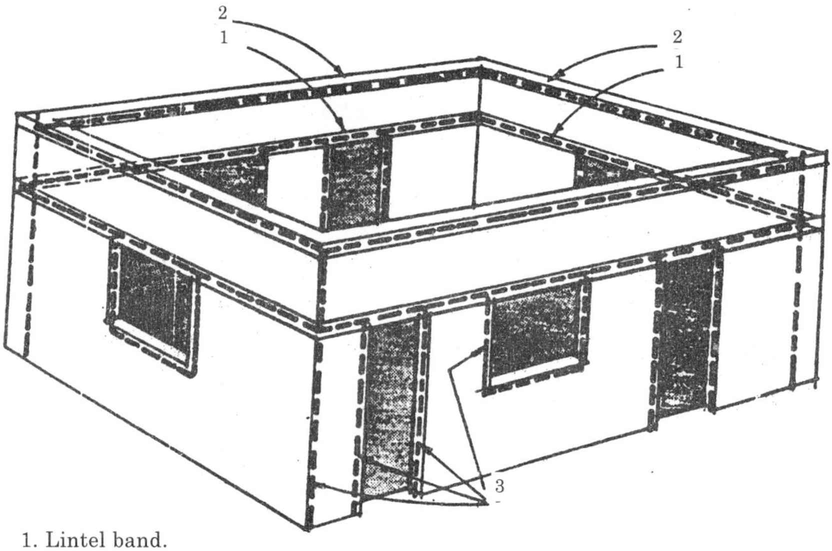

1. Lintel band.
2. Roof band (only for pitched roofs and under flexible roofs and floors)
3. Vertical steel.

Fig. 4.9. Arrangement of reinforcing masonry buildings.

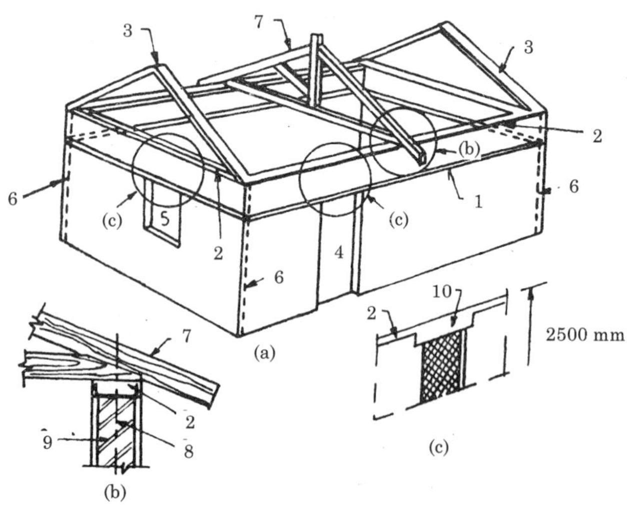

1. Lintel band
2. Eave level (roof) band
3. Cable band
4. Door
5. Window
6. Vertical steel bar
7. Rafter
8. Holding down bolt
9. Brick/Stone wall
10. Door lintel integrated with roof band
    (*a*) Perspective view
    (*b*) Details of truss connection with wall
    (*c*) Detail of integrating door lintel with roof band

Fig. 4.10. Arrangement of reinforcing masonry building having pitched roof.

SECTION-2

4.11. INTRODUCTION TO IS 13828 : 1993

IS : 13828, 1993 is the Indian Standard Guidelines for improving earthquake resistance of low strength Masonry buildings. This masonry includes burnt clay brick or stone masonry in weak mortars, like clay, mud etc. These standards are applicable in Seismic zones III, IV and V. As explained in the last chapter, these constructions are non-engineered and cannot be made free from collapse under severe seismic shaking (MMI VIII or more). The guidelines given in the code may only increase seismic resistance and reduce chances of collapse. The various codal provisions for improving earthquake resistance of low strength masonry buildings are explained below :

4.11.1. Earthquake Resistance Features of Stone Masonry

Low strength stone masonry buildings are weak against earthquakes and should be avoided in high seismic zones. The various features for improving the earthquake resistance of stone masonry buildings are explained below :

(1) General Construction Aspects :

- The wall thickness should not exceed 450 mm, preferably 350 mm.
- The height of a storey should not be greater than 3.0 m.
- In general, the stone masonry buildings should not be taller than 2 storeys when built in cement mortar and 1 storey when built in lime or mud mortar.
- The unsupported length of wall between cross walls should be limited to 5 m. For longer walls, cross supports, called buttresses should be at spacing not more than 4 m.

(2) Mortar :

- Clay mud mortar should be avoided.
- Cement-sand mortar 1 : 6 (or richer) and lime sand mortar 1 : 3 (or richer) should be used for building construction.

(3) Wall Construction :

- The masonry walls shoujld be built in lifts not more than 600 mm.
- **Through stones :** Inner and outer faces of the wall should be bonded with "through stones" (Over full thickness of wall). Through stones should be used in every 600 mm lift at not more than 1.2 m apart horizontally as shown in Fig. 4.11.
- If full length "through stones" are not available then band stones in pairs, (each about 3/4 of wall thickness) should be provided. (Fig. 4.11)

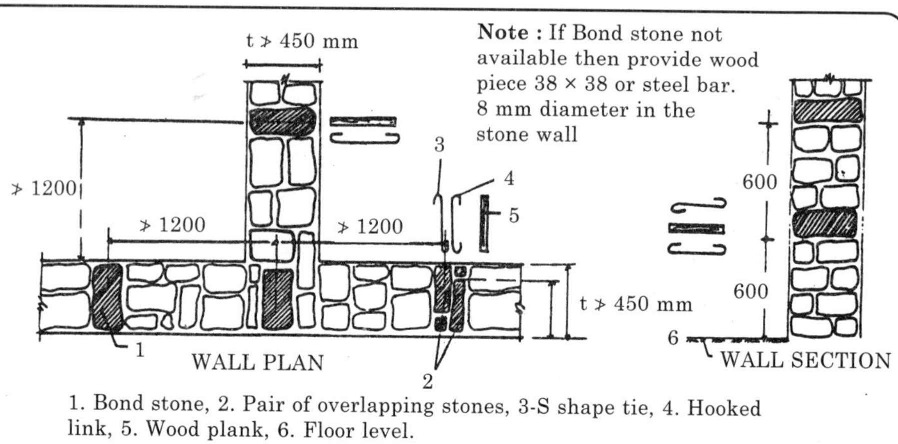

Fig. 4.11. Use of through stones.

- In place of "through stones" steel bars of 8 to 10 mm diameter in S shape or a hook may be used with a cover of 25 mm from each face (Fig. 4.11).
- Wooden bars of 38 mm × 38 mm may be used in place of through stones. The wood used should be well preserved.
- Use of long stones should also be made at corners and junctions of wall to break the vertical joint and for bonding perpendicular walls.

(4) Opening in Walls : These are same as in case of masonry buildings and explained in art. 4.9.

- The recommended opening criteria are shown in Fig. 4.12.
- Ventilator "where used" may be made 450 × 450 mm or smaller.

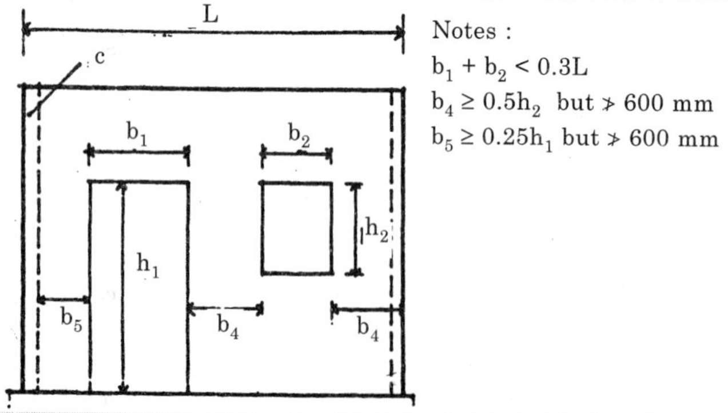

Fig. 4.12. Openings sizes in stone masonry building.

(5) Horizontal Bands : All the horizontal reinforcing arrangement for buildings in art. may be used for random rubble stone masonry construction also.

- The stone masonry buildings must have horizontal bands. These bands can be made of wood or reinforced concrete.
- It is important to provide at least one band (either lintel band or roof band) in stone masonry construction.

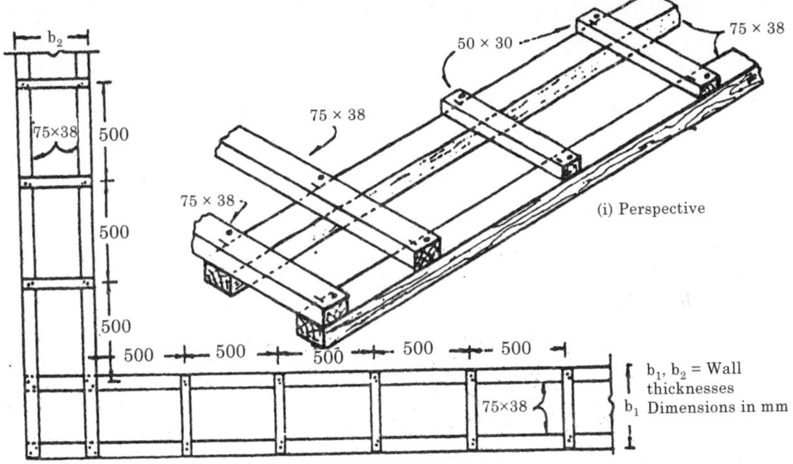

Fig. 4.13. Lintel level wooden band on all load bearing walls.

(6) Vertical Reinforcement in Walls : Vertical steel should be provided at the corners, T junctions of stone walls and at jambs of openings as per Table 4.4.

TABLE 4.4. Recommended Vertical steel in stone masonry buildings.

No. of storeys category	Diameter of mild steel single bar in mm at each critical section B	C	D
One	14	16	20
Two	16	–	–

Two storeyed building with random rubble stone, masonry are not recommended in categories C and D.

The arrangement of vertical steel in random rubble masonry is shown in Fig. 4.14.

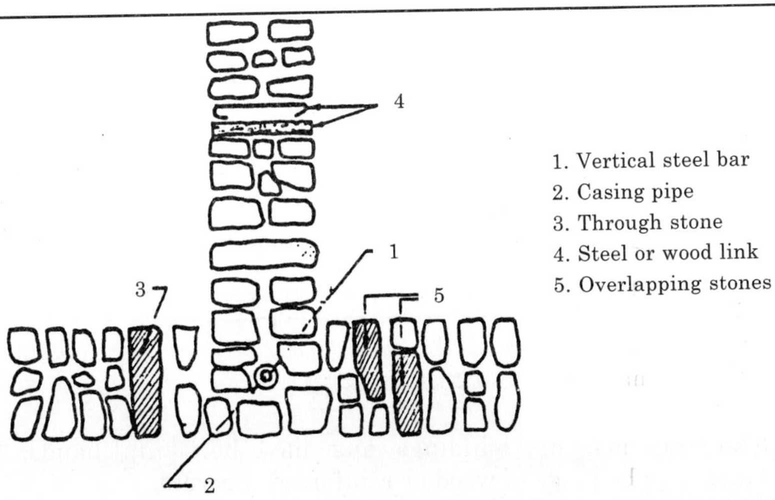

Fig. 4.14. Vertical steel in stone masonry.

Note : (1) Introduction of IS 1893:2016 (Part 1) Indian Standard for Earthquake resistant design of strucutre is given in chapter 7.

(2) IS 13920:2016, Indian Standard code of practice for Ductile Detailing of Reinforced Concrete Structures subjected to Seismic forces is discussed in chapter 6.

TERMS TO REMEMBER

1. **Strengthening :** The measures used to upgrade the earthquake resistance of a damaged building to make it safe for future earthquakes.
2. **Retrofitting :** To upgrade the earthquake resistance of existing building (which are in use) to the level of present day codes.
3. **Ductility :** It is the property of a material to undergo large elongation before breaking.
4. **Deformability :** Deformability is the ability of a structure to undergo large deformations without collapse.
5. **Damageability :** It refers to the ability of a structure to undergo substantial damage without partial or total collapse.
6. **Moment Resistant Frame :** A space frame capable of carrying all vertical and horizontal loads by developing bending moments in the members and at joints.
7. **Load Bearing Wall Construction :** It consists of masonry, concrete or reinforced concrete walls which support vertical and horizontal loads.
8. **Band :** A band is reinforced concrete or reinforced brick runner provided in the walls to tie them together and to impart bending strength to them.
9. **Dowel bars :** These are the U shape steel bars provided at corners and T junctions of walls to improve the box-action.
10. **Through stone :** Large stone which are over full thickness of the wall.
11. **Adobe :** The sun-dried clay bricks or blocks used for construction of earthen houses.
12. **Box action :** A load bearing wall structure without a frame, which are very well connected so that they act as one unit.

EXERCISES

A. Objective Type Questions :

1. Traditionally built constructions are
   (*a*) Non-engineered (*b*) Engineered
   (*c*) Designed (*d*) None of the above

2. The essential properties of an earthquake resistant construction are
   (*a*) Ductility (*b*) Damageability
   (*c*) Deformability (*d*) All of the above
3. Traditional masonry constructions have
   (*a*) High tensile and high shear strength
   (*b*) Low tensile and low shear strength
   (*c*) High tensile and low shear strength
   (*d*) Low tensile and high shear strength
4. Collapse of the masonry structures causes heavy loss of lives because of
   (*a*) Low strength (*b*) Large deformations
   (*c*) Heavy mass (*d*) None of the above.
5. Masonry constructions are
   (*a*) Load bearing (*b*) Rigid frames
   (*c*) Moment resistant frames (*d*) None of the above.
6. Separation of a large building into smaller blocks can lead to
   (*a*) Strong building (*b*) Weak building
   (*c*) More ductility (*d*) Symmetry and regularity
7. A good earthquake resistance building configuration means
   (*a*) Simple plan (*b*) Symmetry
   (*c*) Regularity (*d*) All of the above
8. For earthquake resistant construction, building length should not be more than
   (*a*) Two times its width (*b*) Four times its width
   (*c*) Three times its width (*d*) Its width
9. For large buildings to be constructed on loose soils. The kind of foundation preferred is
   (*a*) Strip footing (*b*) Raft footing
   (*c*) Isolated footing (*d*) None of the above
10. The roof covering should be made of
   (*a*) Light material (*b*) Heavy material
   (*c*) Loose units (*d*) Ductile material
11. As per IS 4326 : 2013 the crushing strength of masonry unit should not be less than
   (*a*) 35 MPa (*b*) 0.35 MPa
   (*c*) 2.5 MPa (*d*) 3.5 MPa
12. Minimum recommended cement : sand mortar for earthquake resistant masonry construction is
   (*a*) 1 : 8 (*b*) 1 : 6
   (*c*) 1 : 4 (*d*) 1 : 2
13. The thickness of load bearing walls should not be less than
   (*a*) 150 mm (*b*) 180 mm
   (*c*) 160 mm (*d*) 190 mm
14. The masonry bearing walls can be made upto
   (*a*) Five storeys (*b*) Three storey
   (*c*) Four storey (*d*) Two storeys.

15. Openings in any storey should be
   (*a*) Scattered at different levels (*b*) At the same level
   (*c*) Near the corner (*d*) Near the centre
16. Bands are provided in masonry buildings. If they increase
   (*a*) Ductility of structure (*b*) Tensile strength
   (*c*) Horizontal bending strength (*d*) None of the above.
17. The loading bearing walls should have
   (*a*) Lintel band (*b*) Plinth band
   (*c*) Roof band (*d*) All of the above
18. Minimum depth of band in masonry construction is
   (*a*) 75 mm (*b*) 50 mm
   (*c*) 60 mm (*d*) 85 mm
19. Grade of concrete used for bands is
   (*a*) M10 (*b*) M20
   (*c*) M15 (*d*) M25
20. The shape of dowel bars provided in masonry construction is
   (*a*)  (*b*) 
   (*c*)  (*d*) None of the above

B. Fill in the Blanks :

1. Masonry construction have very ________ seismic resistance.
2. ________ and ________ type of construction is usually adopted in masonry buildings.
3. ________ codes define methods and techniques which help in planning, designing, detailing and constructing the earthquake resistant structures.
4. Heavier structure means ________ inertia forces.
5. A ________ weight building has better seismic resistance.
6. ________ of a large building into smaller blocks can make a building symmetrical and regular.
7. The building should not be constructed on ________ soils.
8. For large buildings, ________ or ________ foundation is best suited.
9. IS : 4326 : 2013 is the code of practice for earthquake resistant ________.
10. IS : 13827, 2013 deads with the guidelines for improving earthquake resistance of ________ buildings.
11. As per IS : 4326 : 2013, the bricks used in masonry should have crushing strength not less than ________.
12. The mortar used in earthquake resistant buildings should be ________.
13. The thickness of load bearing walls should not be less than ________.
14. ________ must be provided in all load bearing walls and cross walls.
15. ________ band is provided where soil is soft or uneven.
16. The Band should be made of ________ concrete grade.
17. ________ bars are provided to improve the box action.
18. Timber construction should not be greater than ________ storeys.
19. Low strength stone masonry buildings should not be constructed in ________ zones.
20. In stone masonry, ________ stones should be used in walls at not more than 1.2 m apart.

21. In stone masonry, ________ bands must be provided either at lintel or roof level.
22. The roof covering should be made of ________ material.
23. ________ shape of dowel bars are used in masonry construction.

C. State True/False :

1. Framed construction is also called as load bearing walls.
2. A building should be as light as possible, especially the roofs and other storeys.
3. It is advisable to have one long room rather than separate small rooms.
4. Loose soil result in unequal settlement of building.
5. The masonry bearing walls can be built upto a maximum of three storeys.
6. The openings in walls should be small and centrally located.
7. Bands are optional in masonry construction.
8. Plinth band helps in reducing the differential settlement of building.
9. The depth of band should not be more than 75 mm.
10. A band is made up of reinforced concrete only.
11. No vertical steel is provided in category 'A' building.
12. Masonry buildings can be strengthened by providing horizontal and vertical bands.
13. IS 13827 : 1993 is the seismic code for ductile detailing of structures.
14. Masonry construction is non-engineered.
15. In place of 'through' stones, S shaped hooks may be used.
16. Less deformability makes the structure more earthquake resistant.
17. Diagonal bracing in the walls improves the seismic resistance of earthen construction.
18. Multistoreyed and Industrial buildings have box type construction.
19. Asymmetrical buildings are subjected to torsion.
20. Building length should not be more than three times its width.
21. As per IS 4326 : 2013, the masonry buildings are categorized in four types A, B, C and D.
22. The length of masonry wall should not be greater than 15 m.
23. The joints in brick masonry construction should be in a straight line.
24. Either vertical or horizontal bands should be provided in buildings.
25. Seismic codes give guidelines for strengthening masonry buildings.

ANSWERS

A. Objective Type Questions :

1. (*a*)	**2.** (*d*)	**3.** (*b*)	**4.** (*c*)	**5.** (*a*)
6. (*d*)	**7.** (*d*)	**8.** (*c*)	**9.** (*b*)	**10.** (*a*)
11. (*d*)	**12.** (*b*)	**13.** (*d*)	**14.** (*c*)	**15.** (*b*)
16. (*c*)	**17.** (*d*)	**18.** (*a*)	**19.** (*c*)	**20.** (*c*)

B. Fill in the Blanks :

1. Less		**2.** Framed, Box	**3.** Seismic
4. large	**5.** light	**6.** Separation	**7.** soft
8. raft, pile	**9.** Design and construction of buildings		**10.** earthen

11. 3.5 MPa	**12.** 1 (cement) : 6 (sand)			**13.** 190 mm
14. Lintel	**15.** Plinth	**16.** M15		**17.** Dowel
18. two	**19.** high seismic			**20.** Through
21. horizontal		**22.** lighter		**23.** lighter
24. Adobe		**25.** U		

C. State True/False :

1. False	**2.** True	**3.** False	**4.** True	**5.** False
6. True	**7.** False	**8.** True	**9.** False	**10.** False
11. True	**12.** True	**13.** False	**14.** True	**15.** True
16. False	**17.** True	**18.** False	**19.** True	**20.** True
21. False	**22.** True	**23.** False	**24.** False	**25.** True

SHORT ANSWER TYPE QUESTIONS

1. Explain the following terms :
   (*a*) Horizontal bands
   (*b*) Dowel bars
   (*c*) Stud wall construction
   (*d*) Through stones
   (*e*) Diagonal bracing
   (*f*) Box action
   (*g*) Ductility
   (*h*) Lintel band
   (*i*) Deformability
   (*j*) Good structural configuration.

2. Enumerate various types of horizontal bands used in masonry buildings.

3. What is a framed construction ?

4. Why should we opt for lighter buildings in earthquake prone area ?

5. Explain the desirable geometric configuration of an earthquake resistant buildings.

6. Why should we avoid to have projections and suspended parts in buildings ?

7. Why is it not desirable to have long buildings ? What is the ratio of length to width of building ?

8. Which type of soil is not suited for construction in seismic areas and why ?

9. What type of bricks and mortar is suitable for construction in earthquake prone areas ?

10. Give specifications of load bearing walls as per IS : 4326 : 2013.

11. What are the guidelines for providing openings in walls ?

12. Why is it necessary to provide plinth band in buildings constructed on soft soils ?

13. What are the guidelines for constructing walls of stone masonry buildings as per IS : 13828 : 1993 ?

14. Which reinforcing bands are recommended for stone masonry construction by IS : 13828 : 1993 ?

15. What types of roof is suitable for buildings in high seismic zones ?

LONG ANSWER TYPE QUESTIONS

1. List and explain the properties of a good earthquake resistant structure.
2. What are the seismic codes ? List the seismic codes published by Bureau of Indian Standard.
3. Explain the weaknesses of masonry construction which are responsible for its low seismic resistance.
4. What are the types of construction usually adopted in buildings ? Explain them.
5. How does the configuration of a building affect its seismic performance ?
6. Explain the general priniciples to be followed for construction of earthquake resistant buildings.
7. Give the categories of buildings as per IS : 4326, 2013.
8. Explain the various guidelines given by IS : 4326 for construction of masonry walls.
9. What is a band ? Why is it necessary to provide bands in masonry constructions ?
10. What are the seismic strengthening arrangements for masonry construction ? Explain them.
11. Explain different types of horizontal bands in their use.
12. What are dowel bars ? Where are they provided ?
13. List the points to be considered while providing vertical reinforcement in masonry walls.
14. Draw a diagram showing total strengthening arrangement in a masonry building with flat/pitched roof.
15. Explain the various features for improving the earthquake resistance of stone masonry building.
16. Why is it necessary to provide "through stone" in stone masonry walls ? Give various guidelines with diagrams for using through stone in stone masonry walls.

❑❑❑

5

CHAPTER

Seismic Performance of Reinforced Concrete Buildings

INSIDE THIS CHAPTER

5.1. INTRODUCTION

Reinforced concrete buildings are very common in towns and cities. Reinforced concrete (RC) consists of concrete and reinforcing steel. Plain concrete has low tensile strength, so, steel bars are embedded into concrete in the tensile zone to take tension. RC structures have more strength than other masonry structures. A typical RC building is made of beams, slabs, columns and walls supported on a foundation. The system comprising of RC columns and connecting beams is called as **RC Frame**. After columns and floors (slabs and beams) are cast and the concrete hardens, vertical spaces between the columns and floors are usually filled in with masonry walls. These masonry walls are called as Infill walls. Normally, Infill walls are not connected to surrounding RC columns and beams. These buildings are generally multistoreyed. Therefore, they should be designed and constructed carefully by taking earthquake forces into consideration. This chapter discusses the effect of earthquake on RC buildings in terms of various types of failure of RC buildings under earthquake motion. The chapter also focusses on commonly observed damages of Reinforced concrete multistoreyed buildings alongwith their causes.

5.2. FLOW OF INERTIA FORCES

Under horizontal shaking of the ground horizontal inertia forces are generated which are proportional to the building mass. Since, most of the building mass is present at the floor levels, inertia forces develop at the floor levels in the building. These lateral inertia forces are transferred by the floor slab to the walls or columns and then to the foundation which disperses them safely to the ground as shown in Fig. 5.1(*a*). The flow of inertia forces should be smooth and continuous through the building. As inertia forces accumulate downwards from the top of tbe building, the lower storeys experience higher forces than upper storeys [Fig. 5.1(*b*)]. Therefore the lower storeys should be designed stronger than the upper storeys.

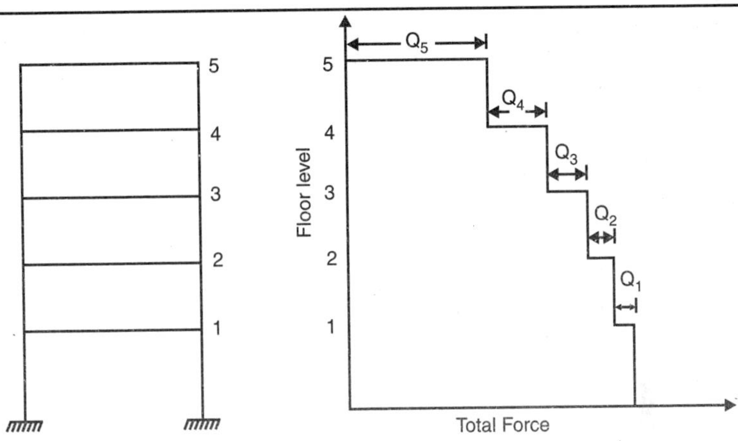

(*a*) Flow of Inertia force. (*b*) Total Lateral Force in a Building.

Fig. 5.1.

5.2.1. Strong Column–Weak Beam Analogy

For a building to remain safe during earthquake shaking, columns (which receive forces from beams) should be stronger than beams, and foundations should be stronger than columns. Further, connections between the beam and columns and foundations should be strong enough to transfer the inertia forces safely to the ground. If columns are made weaker then they suffer severe local damage especially at the joints of lower storey. This damage can lead to collapse of building although the columns at upper storeys are undamaged. On the other hand, if the beams are weaker and columns are stronger then damage will first occur in beams which are ductile enough and cause progressive damage. There will be large deformations (ductile failure) in building before collapse. Therefore in an earthquake resistant building column should be stronger than the beams (Fig. 5.2).

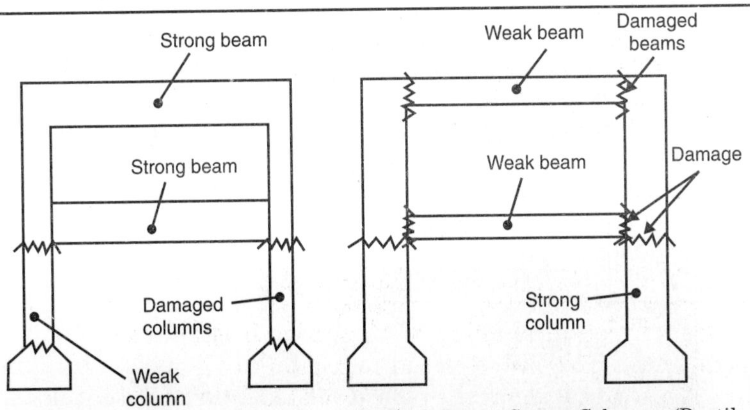

(*a*) Weak Columns Strong Beams (Sudden failure) (*b*) Weak Beams Strong Columns (Ductile failure) (good performance during earthquakes).

Fig. 5.2.

5.3. EFFECT OF IRREGULARITIES ON PERFORMANCE OF RC BUILDINGS

The most important cause of damage of RC buildings during earthquakes is the irregular building configuration. A building that lacks symmetry and has discontinuity in geometry, mass or load resisting elements is called as **irregular building**. The irregularity or discontinuity in a building result in obstruction of flow of inertia forces and cause lot of damage to the building. Similarly, asymmetry of building causes large torsional forces and hence, cause damage to the structure. Therefore, regular and symmetrical buildings should be constructed in earthquake prone areas. The irregularities in a building are two types :

(1) Vertical Irregularities

(2) Horizontal Irregularities

(1) Vertical Irregularities : These irregularities are as a result of sudden change of strength stiffness, geometry and mass over the height of the building. Some of the vertical irregularities are explained below :

(*a*) **Irregularity in Strength and Stiffness :** It is commonly seen that in multistoreyed RC buildings the ground storey is made taller and open. This is done to have space for parking of vehicles or shopping centres/halls etc. This type of RC buildings have only columns in the ground storey and both partition walls and columns in the upper storeys (Fig. 5.3). This configuration results in following :

(*i*) *The ground storey is flexible as compared to upper storeys* which are stiffer because of presence of infill walls. Thus, the upper storeys move as a single unit together and most of the horizontal displacement of the building occur in ground storey (Fig. 5.4). This type of building is called as **Soft Storey Building.** In soft storey buildings, the ground storey columns will be damaged severly if they do not have enough strength and ductility and ultimately lead to collapse of the building.

(*ii*) *The strength of the open ground storey is less as compared to the upper storeys.* Thus, it is also called as weak storey.

Open ground storey buildings are called as Soft Storey Buildings although the ground storey is soft and weak both. Generally, Soft Storey exist at the ground floor but it may exist at other floor levels also.

Fig. 5.3. Soft Storey.

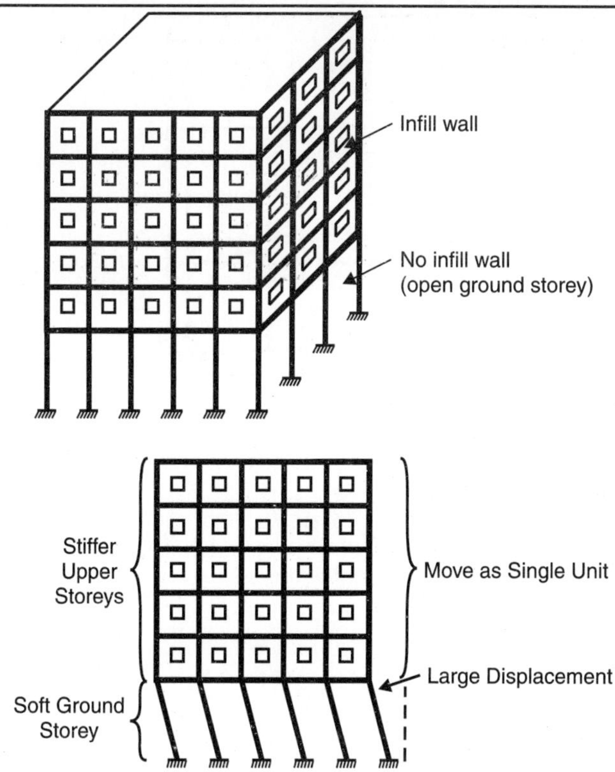

Fig. 5.4. Movement in Open Ground Storey Building.

(*b*) **Vertical Discontinuity :** One of the major cause of damage in RC buildings during strong earthquakes is the discontinuity in the path of load transfer (flow of inertia forces). The structure should have a complete and continuous path for transfer of Seismic forces *i.e.*, floor → beams → columns and walls → foundation → ground. Any discontinuity or irregularity in the path of load transfer result in severe damage and may lead to collapse of the building. Some types of vertical discontinuities are discontinuous columns, discontinuous shear walls or bracings etc. In these, the columns and shear walls do not continue up to the ground but end at an upper level as shown in Fig. 5.5(*a*). This type of configuration is called as **Floating Columns.**

Floating Columns : Most of the buildings have balconies overhanging in the upper storeys to have more area in the upper storeys. Generally in this type of construction, perimeter column of the ground storeys are discontinued and floating columns are provided along the overhanging on the upper storeys of building. These floating columns rest at the tip of the tapered cantilever beams. This type of construction is very dangerous during strong earthquakes and result in damage to the beam column joint and columns as shown in Fig. 5.5(*b*) and may result in total collapse.

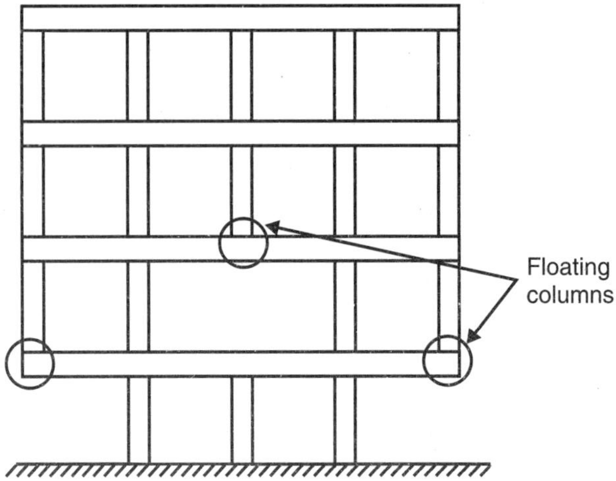

(a) Floating Columns.

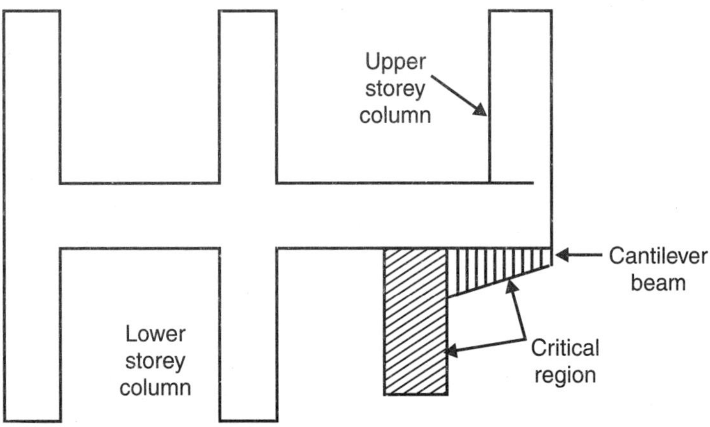

(b) Floating Column Construction.

Fig. 5.5. Floating Column Construction.

(c) **Mass Irregularities :** Mass irregularity is considered to exist when the effective mass of a storey is more than 200% of the effective mass of an adjacent storey. The effective mass of a storey is the dead weight of floor plus partition walls etc. The increased mass result in increased inertia forces and may result in heavy damage. Massive roofs and heavy equipments rooms at upper floors are not advisable as they may induced large inertia forces and cause collapse of the structure during earthquakes. (Refer Fig. 5.6)

(a) Collapsed Building, Earthquake Taiwan, 2016

(b)

Floating Column Failure

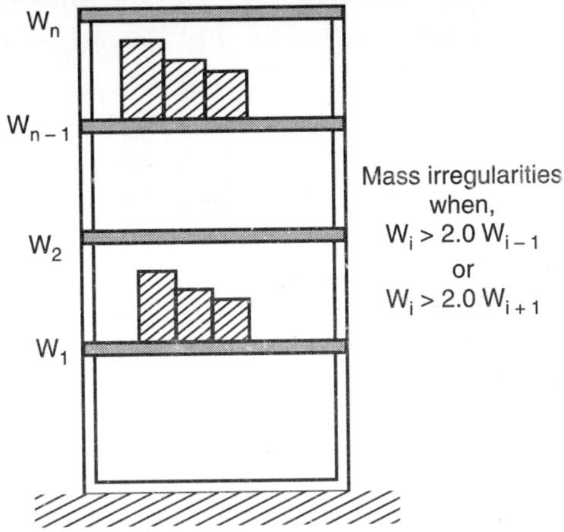

Fig. 5.6. Mass irregularity in Building.

(*d*) **Vertical Geometric Irregularity** : The example of vertical geometric irregularity are shown in Fig. 5.7. When the horizontal dimension of the building in any storey is more than 150% of that in an adjacent storey then the building is said to have vertical geometric irregularity. The solution of this problem is the total separation of the building blocks in plan to make the building regular as already discussed in article 4.7. The separation section should be designed as per the Indian Standard IS 1893 : 2016.

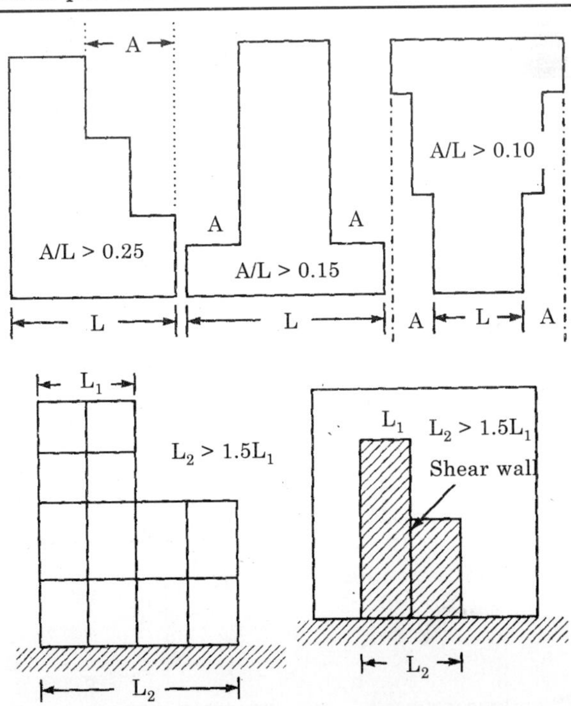

Fig. 5.7. Vertical Geometric Irregularity.

(2) Horizontal Irregularities : These irregularities are caused due to :

(*a*) Asymmetrical plan shapes (*b*) Re-entrant corners

(*c*) Non-parallel system (*d*) Cut-out

(*a*) **Asymmetrical Plan irregularities :** Plan irregularities caused due to asymmetrical plan shapes like L, T, U, F, C etc. which result in large torsional forces. These forces are in addition to the inertia forces and may result in partial or full collapse of the building. Plan irregularities can also be reduced by constructing separate blocks with adequate separation section as discussed in article. Some of the plan irregularities are shown in Fig. 5.8(*a*).

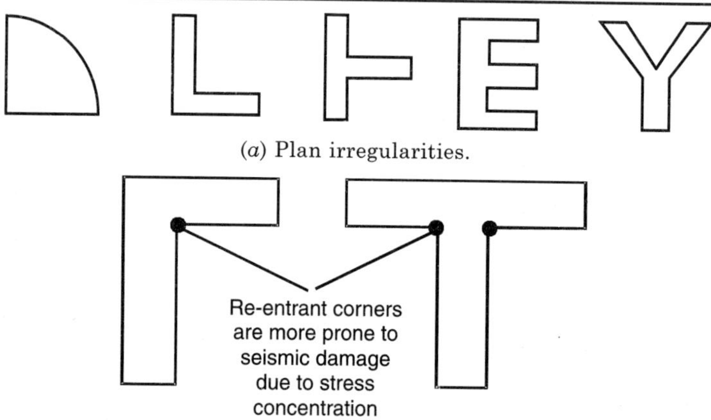

(*a*) Plan irregularities.

(*b*) Re-entant Corners.

Fig. 5.8.

(*b*) *Re-entrant Corners :* The inside corners [Fig. 5.8(*b*)] of an asymmetrical building in plan are subjected to stress concentration during earthquake motion. Thus, these corners are more prone to damage during earthquakes. To avoid this type of damage, either provide a separation joint between two wings of buildings or tie the building together strongly at the common corners.

(*c*) *Non-parallel System :* In the triangular or trapezoidal shaped buildings, the columns or shearwalls are not parallel or symmetrical about the two axis of the building as shown in Fig. 5.9(*a*). This result in high torsional forces and cause more damage to the buildings during earthquake motion. These buildings should be designed carefully taking into account the additional torsional forces.

(*d*) *Diaphragm Discontinuity :* In buildings, roof or floor acts as diaphragms (horizontal resistance element). The diaphragm discontinuity is because of a cut-out or large openings [Fig. 5.9(*b*)]. This causes reduction in the load carrying capacity of diaphragms and may cause damage during earthquakes.

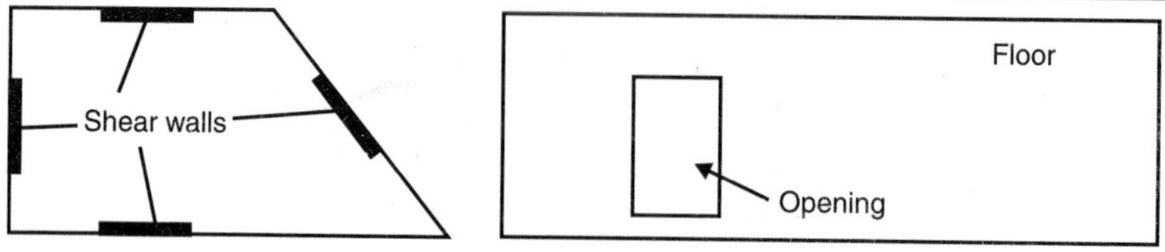

(*a*) Non-parallel System. (*b*) Diaphragm discontinuity.

Fig. 5.9.

5.3.1. Definitions of Irregular Building

IS 1893 (Part 1) : 2016 defines various types of irregularities and these are given in Tables 4 and 5 of the code.

5.4. IDENTIFICATION OF SEISMIC DAMAGES IN REINFORCED CONCRETE BUILDINGS

Reinforced concrete multistoreyed buildings in India were subjected to a strong earthquake in Bhuj, 2001. The major cities affected by this earthquake have RC buildings and the collapse of these buildings resulted in lot of casualties. Damage survey reports of past earthquakes have shown the following main weaknesses in RC frame buildings.

(*a*) Discontinuous/irregular load transfer path.

(*b*) Lack of ductility of structural members.

(*c*) Poor quality of construction materials and faulty construction practices.

These weaknesses have resulted in several types of damages in RC buildings. Common type of damages identified in RC buildings are explained below :

(1) Soft storeys

(2) Floating columns

(3) Plan and mass irregularities

(4) Poor quality of construction material and faulty construction practices

(5) Pounding of adjacent buildings

(6) Inadequate ductile detailing of structural components.

(7) Damage to Non-structural elements

(1) Soft Storey : It is a common practice of constructing open, taller first storey for parking, large halls or banking and meeting room. Due to this, the first storey has less strength and stiffness and result in weak or soft storey. It has been observed from the survey that the damage is due to collapse and buckling of columns of first storey (open storey). In some cases the soft storey has resulted in collapse of full building also (Fig. 5.10).

(*a*) 1971 San Fernando Earthquake.

(*b*) 2001 Bhuj Earthquake.

Fig. 5.10. Soft Storey Failure.

(2) Floating Columns : Most of the buildings in cities have balconies in the upper floors to cover more area. These balconies overhangs upto 1.2 m to 1.5 m beyond the columns of ground storey. In the upper stories, the floating columns are provided which rest at the tip of the tapered overhanging beams. Fig. 5.11 shows damage in reinforced concrete building due to floating columns. This is because of less strength of ground floor columns, failure of cantilever beams and beam-column joints.

Fig. 5.11. Failure Due to Floating Columns.

(3) Plan and Mass Irregularity : The complex shapes (Fig. 5.12) of RC buildings results in irregular buildings which are very dangerous for earthquake prone areas. These irregular building cause discontinuity in the path of transfer of inertia forces. Two buildings of Ahmedabad Mansi complex and Shikhar apartment have irregular shapes and collapsed during the 2001, Bhuj earthquake (Fig. 5.13). Shikhar apartment was U shape and without any separation joint. Mansi complex was C shape (plan irregularity) and had a massive swimming pool at 10th floor (mass irregularity). The heavy mass of swimming pool generated very large inertia forces and resulted in collapse of the building.

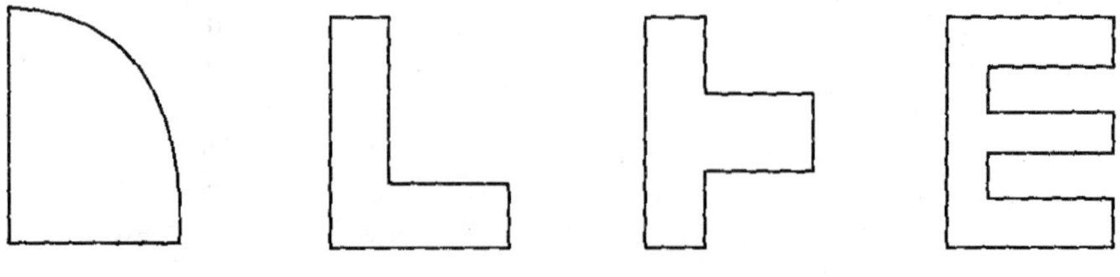

Fig. 5.12. Plan Irregularity.

(a) Total collapse of half portion of irregular building

(b) Collapse of a block of irregular building

Fig. 5.13.

(4) Poor Quality of Construction Material and Corrosion of Reinforcement : Various buildings have failed due to poor quality of construction materials. For example inferior quality of steel or water used result in corrosion of reinforcement which result in spalling of concrete (Fig. 5.14). The corrosion is also caused by insufficient cover, porous concrete and less compaction. Faulty construction practices include the improper proportioning of concrete ingredients, wrong placement of steel bars, in adequate development length and improper splicing etc.

(5) Pounding Damage : Pounding is caused due to hammering of adjacent buildings of different heights. If the floors of adjacent buildings are at different height then they hit against each other during earthquake motion as shown in Fig. 5.15. The columns of the buildings are damaged due to hammering of floors of buildings (Fig. 5.16). This problem arises when buildings are built without proper separation between them. Damage due to pounding may be minimized by providing proper separation between the buildings or by aligning floors of adjacent buildings.

Fig. 5.14. Pounding Effect.

Fig. 5.15. Pounding Damage.

(6) Inadequate Ductile Detailing of Structural Elements : Ductility is an important properly for an earthquake resistant building. The RC buildings are made ductile by detailing the various elements (beam, column and beam column joint) in such a manner that brittle failure is avoided. For example, one of the commonly observed practice of construction is to make weak column and strong beams (art. 5.3.1) which causes damage to the columns and sometimes collapse of the building also. This practice makes the structure brittle. Thus, it is advisable to construct, columns stronger than beams in order to have ductile failure. Structural elements are commonly seen damaged due to improper ductile detailing. Some of the damages of structural elements are explained below :

(*a*) Column :

- *Improper Column Spacing of Lateral Tties* : Improper spacing of lateral ties in columns lead to shear fialure [Fig. 5.16(*a*)]. Shear failure is brittle and must be avoided in columns by providing ties at closer spacing.
- Lateral ties also confine the concrete in columns. Large spacing of ties lead to lack of confinement of concrete and concrete may spall off during earthquakes [Fig. 5.16(*b*)].
- If the ends of the ties are not hooked properly then they may open up during earthquake motion and cause damage to columns.
- *Incorrect Splicing of Column Bars* : This may be because of less lap length provided to the spliced bars or splicing all the bars at same section. This type of failure results in crushing of concrete in the compression zone and buckling of bars also. This type of damage is very serious and may cause collapse also.

Fig. 5.16(*a*) and (*b*). Failure of Columns in RC Building.

(*b*) Beam : Very little cases of beam failure are noticed during past earthquake studies but beam column joint failures are very common in RC buildings. (Fig. 5.17). The main cause of damage of beam column joint is improper detailing of reinforcement at the joint. Some of the flaws of detailing are as follows :

- Inadequate reinforcement in the joint
- Absence of lateral ties
- Presence of column bar splices in the joint
- Smaller column size.

Fig. 5.17. Damage in Beam Column Joint in RC Building.

(7) Damage to Non-structural Elements

(*a*) **Damage to Infill Walls :** Masonry infill walls are used to fill up spaces between the columns and slab. In routine design the infill walls are not considered while analyzing multistoreyed buildings. The infill walls increase the strength and rigidity of the building. Their, non-inclusion in the design may cause failure of multi-storeyed building. This failure is mainly because of cracking of infill walls, short column effect etc. as explained below.

- *Cracking of the Infill Wall :* During earthquake motion. The reinforced concrete frame deform first and the first crack appear on the plaster along the contact of infill wall with frame. As the deformation of the frame becomes larger, the crack penetrate into the masonry and result into separation of infill wall from the frame. These separate infill wall act as short column and fail due to diagonal cracking [Fig. 5.18] shows the cracking and failure of infill wall.

(*a*) Cracking of infill wall.

(*b*) Failure of exterior walls.

Fig. 5.18. Failure of Infill Walls.

- Failure of exterior walls is also commonly observed which is mainly because of out of plane bending as shown in Fig. 5.18.

- *Short Column Effect :* This case arises when their are many closely spaced windows and masonry walls are built upto in partial height. The columns between the walls behave as short columns due to presence of the partial walls as shown in Fig. 5.20(a). These columns are also called captive columns. A captive column is full storey column whose clear height is reduced by the stiffer masonry infill walls. When floor slab moves horizontally during an earthquake, the upper end of the short column or captive column undergo displacements which is same as the regular column. But the stiff masonry infill walls restrict horizontal movement of the lower portion of the short column. Thus, it deforms by the full amount over the short height and attract large horizontal forces which result in severe damage to the column. Fig. 5.19(*a*) shows short column effect.
- Short column effect can also arise in other cases like buildings on sloping ground and buildings with a mezzanine floor. In all the cases short columns have suffered more damage as compared to taller columns of same storey. Fig. 5.19(*b*) and Fig. 5.19(*c*) shows the failure of short column and buildings with short columns.

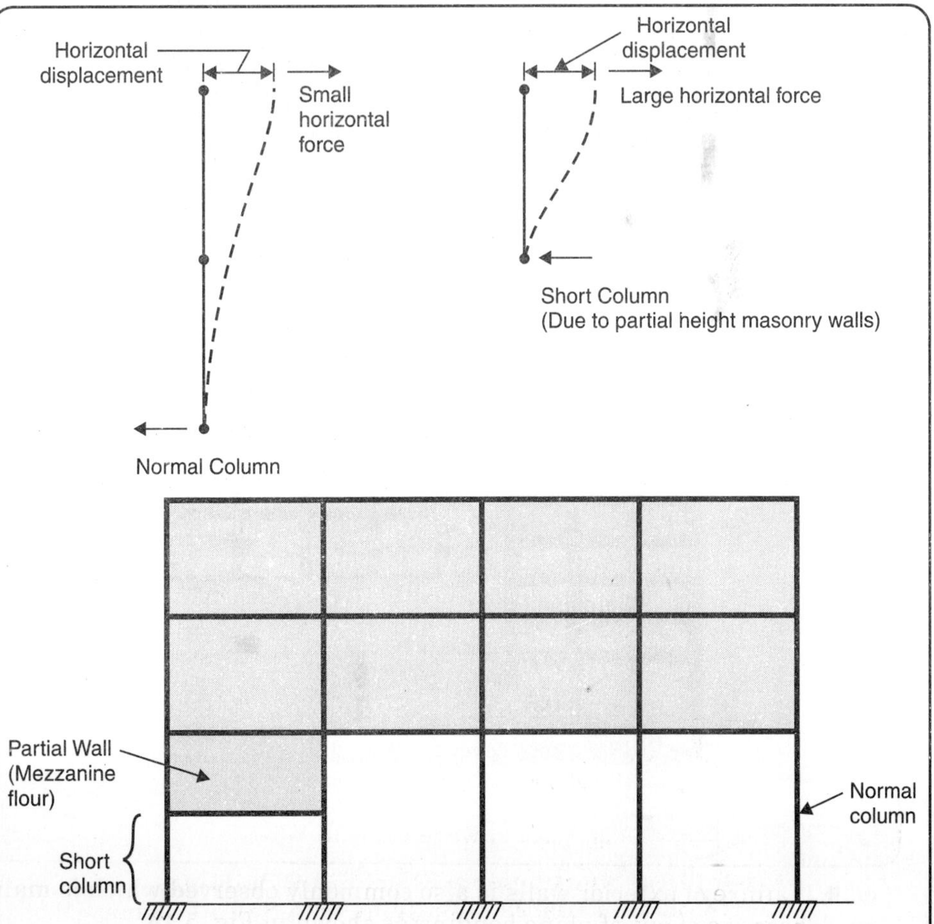

Fig. 5.19(*a*). Short Column Effect.

(*b*)

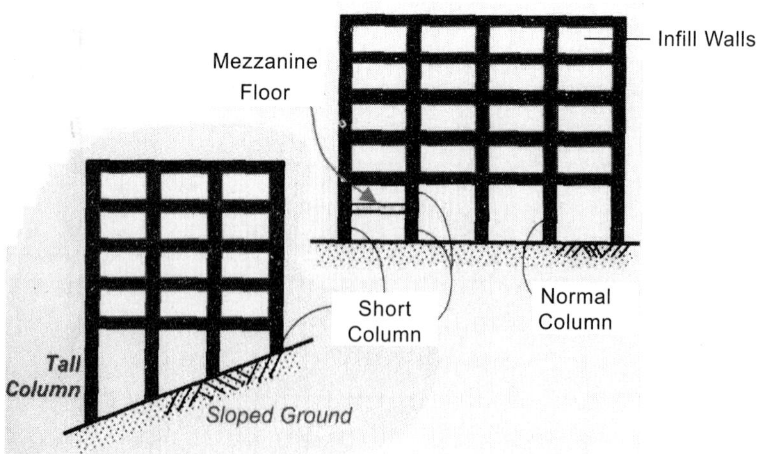

(*c*)

Fig. 5.19. (*a*) Short Column Effect in RC Buildings, (*b*) Short Column Failure, (*c*) Building with Short Column

(*b*) **Damage to Water Tanks and Parapets :** Many times water tanks are constructed at the roof level of the building. These tanks experience large inertia forces due to heavy weight and height of the building. This may lead to damage to the tank. Unreinforced concrete parapets which are not properly tied to the root are also damaged during earthquake.

(*c*) **Damage to Stair Case :** In many multistoreyed buildings, failure of stair case is a major cause of damage. Fig. 5.20 shows the failure of stair cases in RC buildings.

Fig. 5.20. Failure of Stair Cases in RC Buildings.

All the above damages and failure can be prevented upto some extent by adopting earthquake resistant design and detailing of the reinforced concrete buildings. These two aspects are covered in following two codes :

1. IS 13920 : 2016
2. IS 1893(1) : 2016

Hence, study of these codes becomes essential and these are covered in the next two chapters.

Conclusion : The study of damage and their causes of multistoreyed RC buildings lead to the following conclusion :

- The design of RC buildings should be based on Indian Standard Codes IS 1893 (Part 1) : 2016 and IS : 13920 : 2016.
- The irregular buildings (Plan or mass irregularity) should be avoided. In case, they are to be constructed, they should be designed on the basis of dynamic analysis taking the irregularity into consideration.
- The soft storey and floating columns or any other discontinuity in the path of load transfer should be included in the design and dynamic analysis of the building. The detaling should be strictly according to the code IS 13920 in such cases.
- Buildings should be planned as per Strong Column – Weak Beam strategy.
- Quality of construction should be monitored properly and stress should be on good quality of construction and materials.
- Infill walls should be taken into account for structural analysis of the building.

TERMS TO REMEMBER

1. **Irregular Building :** A building which is not symmetrical and has discontinuity or non-uniformity in geometry mass or load resisting elements.
2. **Vertical Irregularity :** The irregularity because of sudden change of strength, stiffness, geometry or mass over the height of the building.
3. **Soft storey :** Open ground storey where in fill walls are not constructed for creating more space, and only columns are there.
4. **Floating column :** The columns which do not continue upto the ground but end at an upper level.
5. **Mass irregularity :** In a multistoreyed building if the effective mass of any storey is more than double of its adjacent storeys.
6. **Short column :** If a portion of column is restrained from moving because of presence of walls or magazine floor then it is called as short column.
7. **In fill walls :** The masonry walls used to fill spaces between the floors and columns.
8. **Pounding :** The hitting or hammering of two adjacent buildings or blocks of a building.

EXERCISES

A. Objective Type Questions :

1. Masonry walls used to fill spaces between columns and floors are called as
   (*a*) In built walls (*b*) Infill walls
   (*c*) Filled walls (*d*) None of the above
2. For earthquake resistant buildings they should have
   (*a*) Strong columns–Strong beams
   (*b*) Weak columns–Weak beams
   (*c*) Strong columns–Weak beams
   (*d*) Weak columns–Strong beams
3. Strong Column–Weak beam design leads to
   (*a*) Sudden failure (*b*) Strong building
   (*c*) Ductile failure (*d*) None of the above
4. A discontinuity in the path of load transfer arise due to
   (*a*) Asymmetry (*b*) Irregularity in mass
   (*c*) Irregularity in geometry (*d*) All of the above
5. Soft storey may lead to failure of building during earthquake because it has
   (*a*) More flexibility and less strength
   (*b*) Less flexibility and less strength

(*c*) No flexibility and less strength
(*d*) None of the above

6. Open ground storey building is called as
(*a*) Open storey (*b*) Flexible storey
(*c*) Soft storey (*d*) Strong storey

7. In an open ground storey building, large displacements occur in
(*a*) Upper storeys (*b*) All storeys
(*c*) Ground storey (*d*) None of the above

8. Discontinuous columns are called as
(*a*) Floating columns (*b*) Flying columns
(*c*) Flowing columns (*d*) None of the above

9. Floating columns cause lot of damage during earthquake because they cause
(*a*) Large deformations in the building
(*b*) Make the building stiffer
(*c*) Discontinuity in path of transfer of inertia forces
(*d*) None of the above.

10. Mass irregularity is considered to exist when the mass of a storey is
(*a*) More than double the mass of adjacent storey
(*b*) Equal to the mass of adjacent storey
(*c*) More than half the mass of adjacent storey
(*d*) All of the above

11. An important property which an earthquake resistant building mus have is
(*a*) Ductility (*b*) Flexibility
(*c*) Brittleness (*d*) All of the above

12. Floating columns are an example of
(*a*) Vertical irregularity (*b*) Horizontal irregularity
(*c*) Short columns (*d*) Plan irregularity

13. A captive column is called as
(*a*) Short column (*b*) Floating column
(*c*) Strong column (*d*) Long column

14. Construction of mezzannine floor or many closely spaced windows with partial built walls cause
(*a*) Strong column effect (*b*) Stiff column effect
(*c*) Short column effect (*d*) Long column

15. Masonry infill walls make the building
(*a*) More ductile (*b*) Less strength
(*c*) More rigid (*d*) None of the above

16. Short column attracts
(*a*) No inertia force (*b*) More inertia force
(*c*) Less Inertia force
(*d*) Force equal to that which other column experience

17. Presence of irregularities cause the following during earthquake
(*a*) No effect
(*b*) Large deformations in building

(c) Make the building stiffer
(d) Discontinuity in the path of load transfer.

18. Re-entrant corners lead to
(a) More rigidity (b) Less strength
(c) More flexibility (d) Stress concentration

19. IS : 13920 : 1993 deals with the
(a) Strength consideration (b) Ductility considerations
(c) Strength and ductility (d) None of the above

20. Steel bars in RCC structures are embedded in the
(a) Tension zone (b) Compression zone
(c) Neutral axis (d) Any where

B. Fill in the Blanks :

1. The masonry walls in reinforced concrete buildings are called as ________.
2. For a good earthquake resistant building, columns should be ________ than beams.
3. ________ building has discontinuity in geometry, mass or load resisting elements.
4. Discontinuity in a building means obstruction in the flow of ________ forces.
5. Weak storey is also called as ________ storey.
6. Open ground storeys (without infill walls) are made to have space for ________.
7. Floating columns are an example of ________ discontinuity.
8. Asymmetric plan of the building makes it ________ irregular.
9. A cutout or large opening in the roof or floor is called as ________ discontinuity.
10. Soft storey is an example of ________ irregularity.
11. An open soft storey has less ________ and ________.
12. Overhanging balconies in upper floors to cover more area have ________ columns.
13. Too close adjacent buildings of different heights can cause ________ during earthquake.
14. Earthquake damage in RCC buildings generally occur due to ________ in horizontal and vertical planes.
15. The open ground storey without infill walls is termed as ________.
16. The columns which don't have foundation are called as ________ columns.
17. Ductility helps in ________ damage due to earthquake.
18. Infill walls ________ the stiffness of building.
19. Improper spacing of lateral ties in columns lead to ________failure.
20. Overlapping of bars is called as ________.
21. Short columns are also called as ________ column.
22. Partial height infill walls result in ________ column effect.
23. Short column attract ________ horizontal force.
24., Re-entrants corners are subjected to ________ concentration.

25. Mass irregularity exist when the effective mass of a storey is more than _________ of the effective mass of an adjacent storey.

C. State True/False :

1. Reinforced concrete is good in tension as well as compression.
2. Upper storeys should be designed stronger than the lower storeys.
3. The weak columns and strong beams in a building may result in sudden failure.
4. Irregular buildings are subjected to more damage during earthquake.
5. Open ground storey is flexible as compared to upper storeys.
6. Vertical discontinuity do not affect the flow of inertia forces.
7. Open ground storey building is called as soft storey buildings.
8. Floating column is an example of horizontal discontinuity.
9. The vertical geometric irregularity can be tackled by separating building blocks.
10. Columns are also called as diaphragms.
11. Diaphragm discontinuity causes reduction the load carrying capacity.
12. Infill walls increases the flexibility of the building.
13. Soft storeys make the structure more earthquake resistant.
14. Load should have clear path from roof to the foundation.
15. Infill walls should be taken into consideration while designing RC builgings.
16. Shear walls in buildings should not be placed parallel to each other.
17. Asymmetric plan of a building causes more torsional forces.
18. Floating columns are desirable for earthquake resistant construction.
19. Large spacing of lateral ties result in diagonal cracks.
20. Presence of columns of different heights in a building give rise to short column effect.
21. Ductility decreases the earthquake resistance of a structure.
22. Inadequate ductile detailing makes the structure brittle.
23. Pounding can be avoided by making floor of the adjacent buildings at different level.
24. Improper joint reinforcement can result in more damage to beam column joint.
25. Infill walls act as short column during earthquake.

ANSWERS

A. Objective Type Questions :

1. (*b*)	2. (*c*)	3. (*c*)	4. (*d*)	5. (*a*)
6. (*c*)	7. (*c*)	8. (*a*)	9. (*c*)	10. (*a*)
11. (*a*)	12. (*a*)	13. (*a*)	14. (*c*)	15. (*c*)
16. (*b*)	17. (*d*)	18. (*d*)	19. (*b*)	20. (*a*)

B. Fill in the Blanks :

1. Infill walls	**2.** stronger	**3.** Irregular	**4.** Inertia
5. Soft	**6.** parking	**7.** Vertical	**8.** Horizontally
9. Diaphragm	**10.** Vertical	**11.** Strength, stiffness	
12. Floating	**13.** Pounding	**14.** Irregularity	
15. Soft	**16.** Floating	**17.** reducing	**18.** increases
19. Shear	**20.** Splicing	**21.** Captive	**22.** Short
23. larger	**24.** stress	**25.** 200%	

C. State True/False :

1. True	**2.** False	**3.** True	**4.** True	**5.** True
6. False	**7.** True	**8.** False	**9.** True	**10.** False
11. True	**12.** False	**13.** False	**14.** True	**15.** True
16. False	**17.** True	**18.** False	**19.** True	**20.** True
21. False	**22.** True	**23.** False	**24.** True	**25.** True

SHORT ANSWER TYPE QUESTIONS

1. Explain the following terms :
   (*a*) Soft storey
   (*b*) Floating columns
   (*c*) Short column effect.
   (*d*) Diaphragm discontinuity
   (*e*) Pounding
   (*f*) Infill walls

2. What do you understand by the term irregular building ?

3. Define the term vertical irregularity and horizontal irregularity.

4. What is mass irregularity ?

5. Why it is not good to have closely spaced windows or partial height masonry walls.

6. What is meant by captive column.

7. List any three common types of damages identified in RC buildings.

8. Why re-entrant corners are not desirable for earthquakes resistant construction ?

9. How can we reduce the effect of vertical and plan irregularities ?

10. Why is it not advisable to construct massive roofs and heavy equipment rooms ?

LONG ANSWER TYPE QUESTIONS

1. Explain the flow of inertia forces in a building. How does the inertia force varies along the height of building.

2. Why is strong columns–weak beams configuration is preferred for making earthquake resistant buildings.

3. What is an irregular building ? How does the irregularity affect the seismic performance of any building.

4. What is soft storey ? Why is the soft storey construction dangerous in earthquake prone areas.
5. What is floating column. How does their presence affect the seismic performance of a building.
6. What are the various types of irregularities. Explain the different types of horizontal irregularities.
7. What are the main weaknesses of RC frame buildings.
8. What are the common types of damages observed in RC buildings.
9. What are the damages caused due to in adequate ductile detailing of structural components.
10. Explain the commonly observed damage caused to non-structural components during earthquakes.
11. What is the short column effect. Explain the conditions/constructions resulting in formation of short columns.
12. Explain various types of commonly observed damages of RC buildings during earthquakes and their causes also.

❑❑❑

6

CHAPTER

Ductile Detailing of Reinforced Concrete Buildings (IS 13920:2016)

INSIDE THIS CHAPTER

6.1. INTRODUCTION

One of the most important property which an earthquake resistant structure must have is ductility. A ductile material is the one which can undergo large elongations (strains) while resisting loads. When applied to reinforced concrete members and structures, ductility means the ability to sustain significant inelastic deformations before collapse. A brittle material fails suddenly while ductile material gives sufficient warning before collapse thus saving many lives. It is very important to incorporate ductility into the structures to make them earthquake resistant. To have sufficient ductility, the designer should pay attention to detailing of reinforcement. Sufficient amount of ductility can be ensured by following simple design details.

Provision for ductile detailing in the members of reinforced concrete buildings are given in IS : 13920 : 2016 titled "Ductile Design and Detailing of Reinforced Concrete Structures, Subjected to Seismic Forces". These provisions are for the anchorage, splices, bar cut offs and joint details etc. It is observed in past earthquakes that the problems in structural detailing may also be one of the cause of damage. Thus, this chapter focusses on the provisions of ductile detailing for RC buildings and important clauses of IS13920 : 2016 to understand the importance of ductility of reinforced concrete structures subjected to seismic forces.

6.2. CODAL PROVISION OF IS 13920 : 2016

(General Specifications (Clause 5.0))

- The design and construction of reinforced concrete buildings should be as per IS : 456 : 2000 except as modified by this code.
- Minimum grade of structural concrete is M20.
- For all buildings which are more than 15 m in height, the minimum grade of concrete used shall be M25 for seismic zones III, IV and V.
- Steel reinforcement of grade Fe415 or less shall be used. Strong steel is not preferred in earthquake prone region because low strength steel are more ductile, show long yield and greater breaking strain. Higher strength steel is good for other structural elements where possibility of damage is less under earthquake loads.
- High strength deformed bars (TMT) having elongation more than 14.5% may be used.

6.3. FLEXURAL MEMBERS (CLAUSE 6.0)

The beams and slabs are flexural members which are designed to resist flexure (bending). The cracking pattern of a beam is shown in Fig. 6.1. The vertical cracks at the centre are the bending or flexure cracks. The cracks at

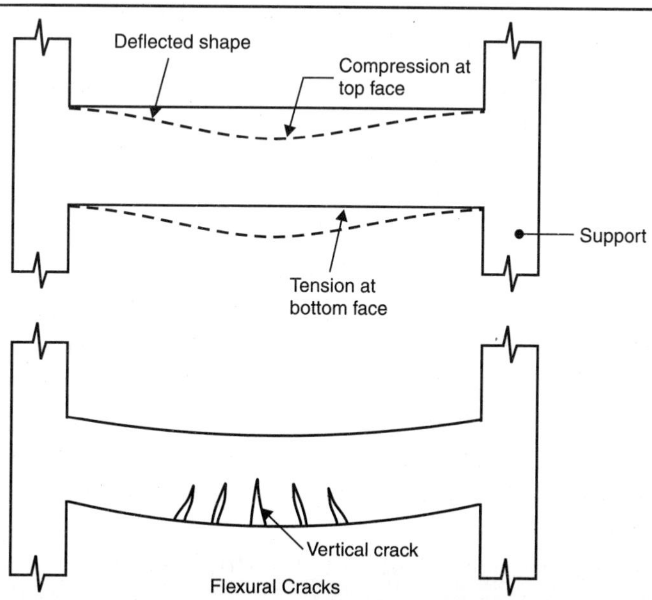

Fig. 6.1. Cracking pattern in a beam.

the end are diagonal tensions cracks caused due to shear. Reinforcement is provided to avoid all types of cracking. The various provisions of reinforcement and detailing for beams are as follows :

- The flexural members are subjected to very little axial stresses. The factored axial stress on these members under earthquake loading shall not exceed 0.08 fck where fck is the characteristic compressive strength of concrete.
- The member should have width to depth ratio of more than 0.3.
- The minimum width of the member is 200 mm.
- The depth of the member should be less than one fourth of the clear span.

6.3.1. Longitudinal Reinforcement

Longitudinal bars are provided in the beams to resist bending cracks. In frames, both top and bottom faces of the beam (at the ends) are subjected to tension, thus steel bars are required on both faces at the ends and on the bottom face at the middle portion of beam. Figure 6.2 shows the critical sections in a beam of a frame where tension may develop during earthquake and steel reinforcement is required.

The codal recommendations for longitudinal reinforcement in a beam are

- At least two 12 mm diameter bars go through the full length of the beam at the top as well as the bottom of the beam.
- At the ends of the beam, the amount of steel provided at the bottom is at least half of that at top.
- The steel provided at each of the top and bottom face of the member at any section shall be equal to one fourth of the maximum negative moment steel provided at the ends of the beam.
- At any exterior joint, the top and bottom bars in a beam should have anchorage length as follows :

  = Development length in tension + 10 ϕ – allowance for 90° bend
- Steel reinforcement bars are available in lengths of 12-14 m, thus, it is necessary to overlap bars. If longer bars are required, such laps of then longitudinal bars should not be made within a distance of 2d from the column face.
- The bars should not be overlapped at the locations where the beam is subjected to greater moments (for example : middle length of beam).
- Lap splices are not provided within a joint.
- The lap length should not be less than the development length of the largest bar in tension.
- At the overlapping lengths, closed links should be provided over all the overlapping (spliced length and at a closer spacing) zone, and the spacing of these links should not exceed 150 mm.
- Not more than 50% of the bars shall be spliced at any section.

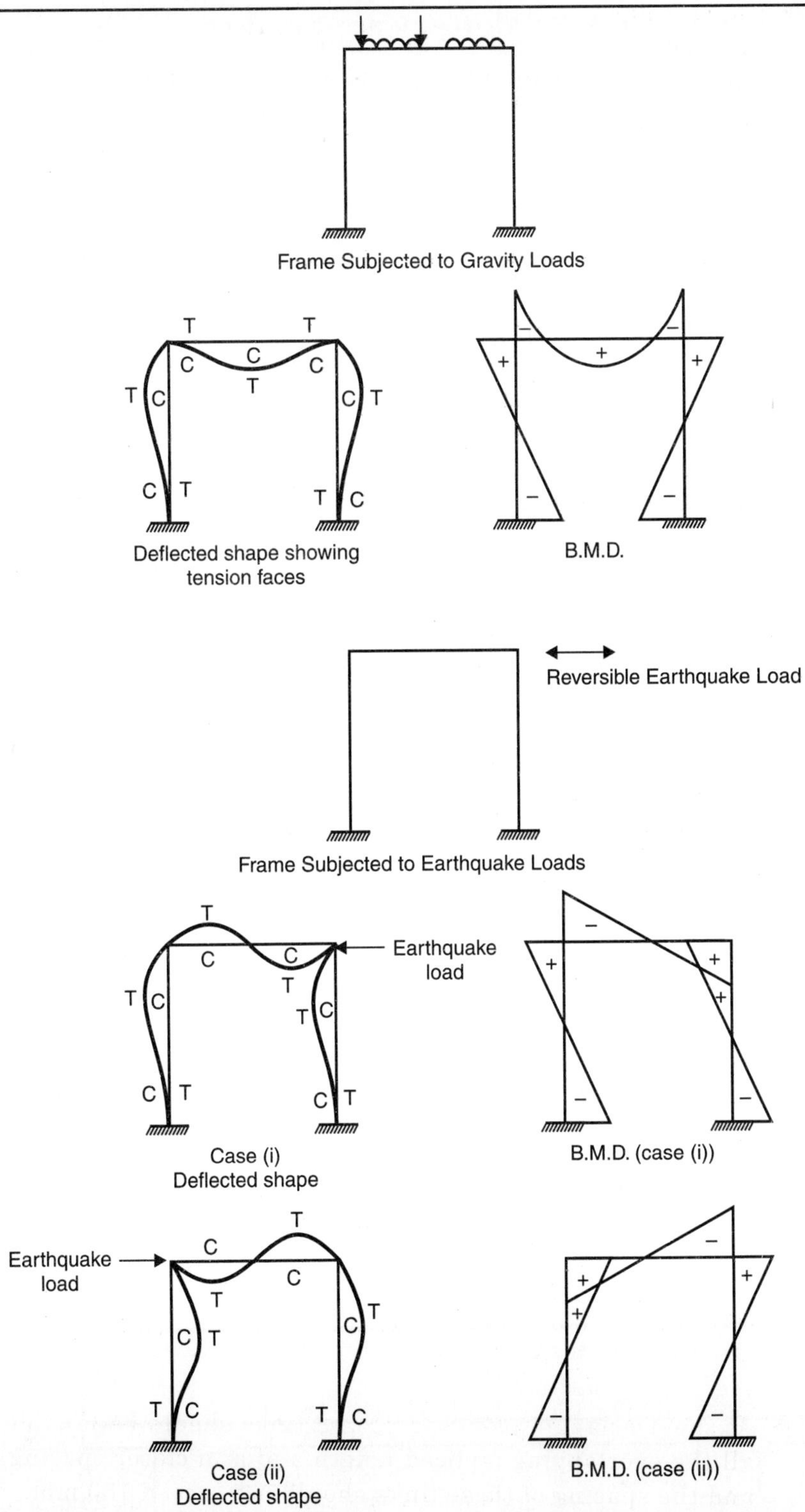

Fig. 6.2. Tension face in beam and columns during earthquake.

6.3.2. Transverse Reinforcement—Stirrups

Stirrups in RC beam has following functions :

(*a*) They carry the vertical shear force and hence resist diagonal shear cracks.

(*b*) They prevent the concrete from bulging outwards.

(*c*) They prevent the buckling of the longitudinal bars.

IS 13920 : 2016 gives following recommendations related to stirrups in reinforced concrete beams (Clause 6.3 of code)

(*a*) Only vertical links are to be used in beams.

(*b*) The diameter of stirrups must be at least 8 mm.

(*c*) Both ends of the vertical stirrups should be bent into a 135° hook and extended sufficiently beyond it to prevent opening of the hook as shown in Fig. 6.3.

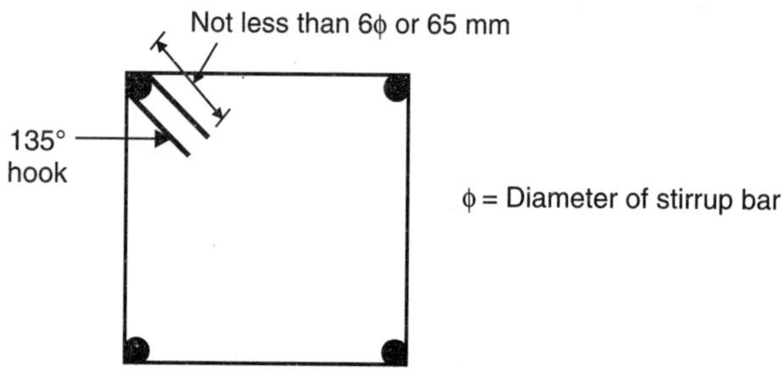

Fig. 6.3. Stirrups in beams.

(*c*) The maximum spacing of stirrups should not exceed half the depth of beam.

(*d*) The contribution of bent up bars, inclined stirrups and concrete should not be considered while calculating shear resistance..

(*e*) The spacing of stirrups over a length of 2d at either end of a beam should not exceed.

(*i*) d/4

(*ii*) 8 times the diameter of the smallest longitudinal bar but in no case greater than 100 mm.

(*f*) The first stirrup should be at a distance not greater than 50 mm from joint face.

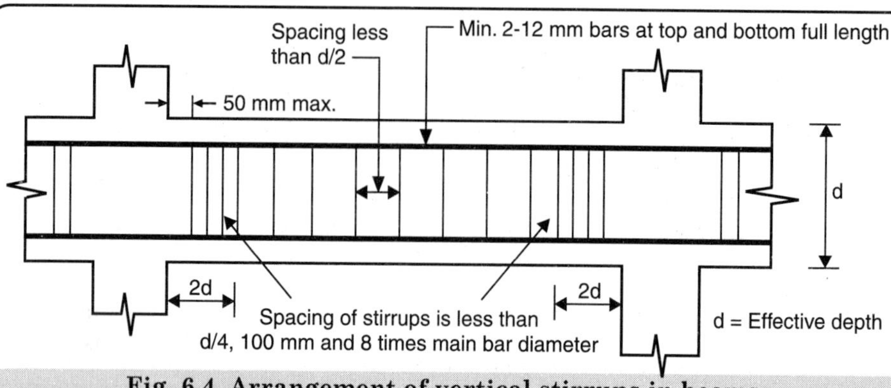

Fig. 6.4. Arrangement of vertical stirrups in beams.

6.4. COLUMNS AND FRAME MEMBERS SUBJECTED TO AXIAL LOAD AND BENDING (CLAUSE 7.0)

Columns are the vertical compression members which are subjected to axial load and bending. They contain two types of reinforcement

(*a*) Longitudinal bars

(*b*) Transverse reinforcement

The earthquake load can cause tension on both faces of columns in frames because of their revsersible nature as shown in Fig. 6.2.

The codal provisions applicable to columns and to frame members which have a factored axial stress > 0.08 fck under the effect of earthquake load are given below :

- The minimum dimension of the member should not be less than 300 mm.
- The ratio of the cross-sectional dimensions (shortest/other dimension), i.e., aspect ratio should not be less than 0.45.

6.4.1. Longitudinal Reinforcement of Columns (Clause 7.3)

In the case of RC columns, it is very common practice to lap columns bars. The lap length depends on types of reinforcement and concrete.

- Lap length should be provided ONLY in the middle half of the column and not within a joint or within 2d distance from face of beam.
- Only half the vertical bars in the column are to be lapped at a time in any storey.
- When laps are provided ties must be provided over the entire lap length at spacing not exceeding 100 mm centre to centre.
- The ductile detailing of a column with spliced bars is shown in Fig. 6.6.
- Lap splices are not used for bars of diameter larger than 32 mm, for them only mechanical spicing is to be adopted.

6.4.2. Transverse Reinforcement of Columns

Transverse reinforcement consisting of closed links in columns also serve three purposes :

(*a*) They carry the horizontal shear forces induced by the earthquake. Thus prevent diagonal cracking.

(*b*) They hold together the vertical bars and keep them in place.

(*c*) They prevent buckling of the vertical bars.

The following are the codal provision for lateral ties of a column (Clause 7.4)

- Transverse reinforcement for circular columns should be spiral or circular.
- Minimum diameter of links is 8 mm for bars up to 32 mm diameter and 10 mm when bars of diameter greater than 32 mm are used.

- For rectangular columns, lateral ties is a closed stirrup having a 135° hook with appropriate extension of 6 times bar diameter but not less than 65 mm as shown in Fig. 6.5.
- The parallel legs of ties should not be more than 300 mm apart. If it is more than 300 mm then a cross tie or a pair of over lapping ties are requied as shown in Fig. 6.5(*b*) and (*c*).
- The spacing of ties should not exceed half the least lateral dimension except where special confining reinforcement is provided.

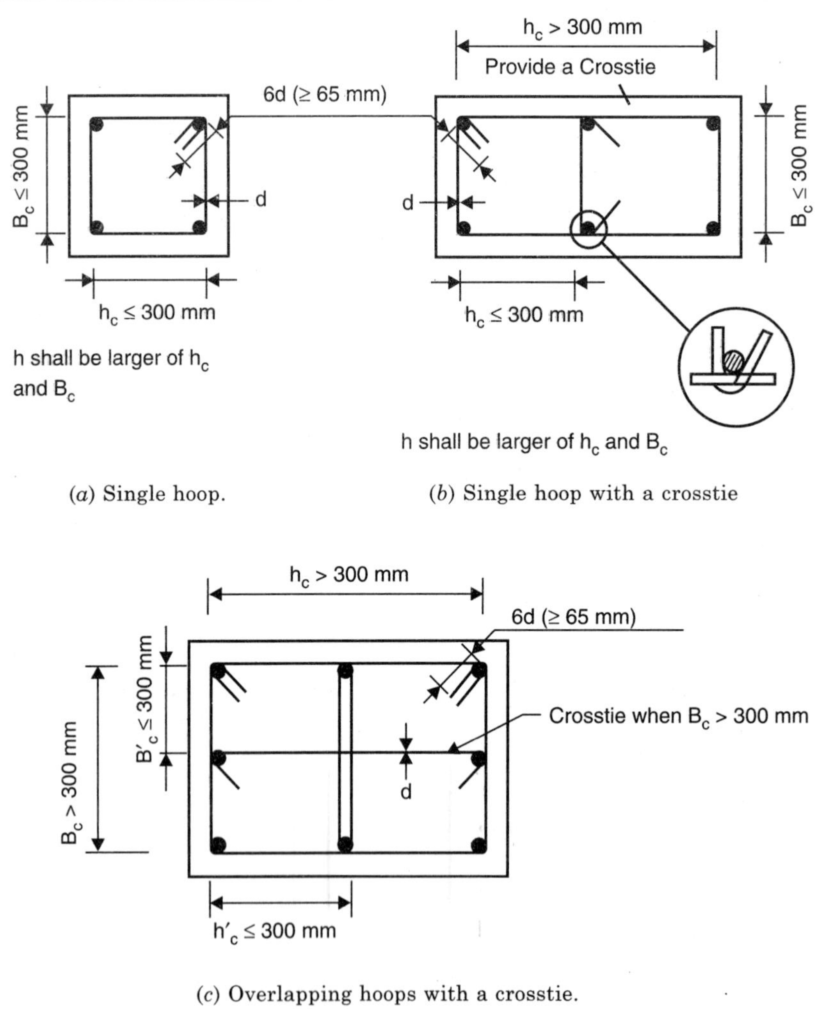

Fig. 6.5. Transverse reinforcement in column.

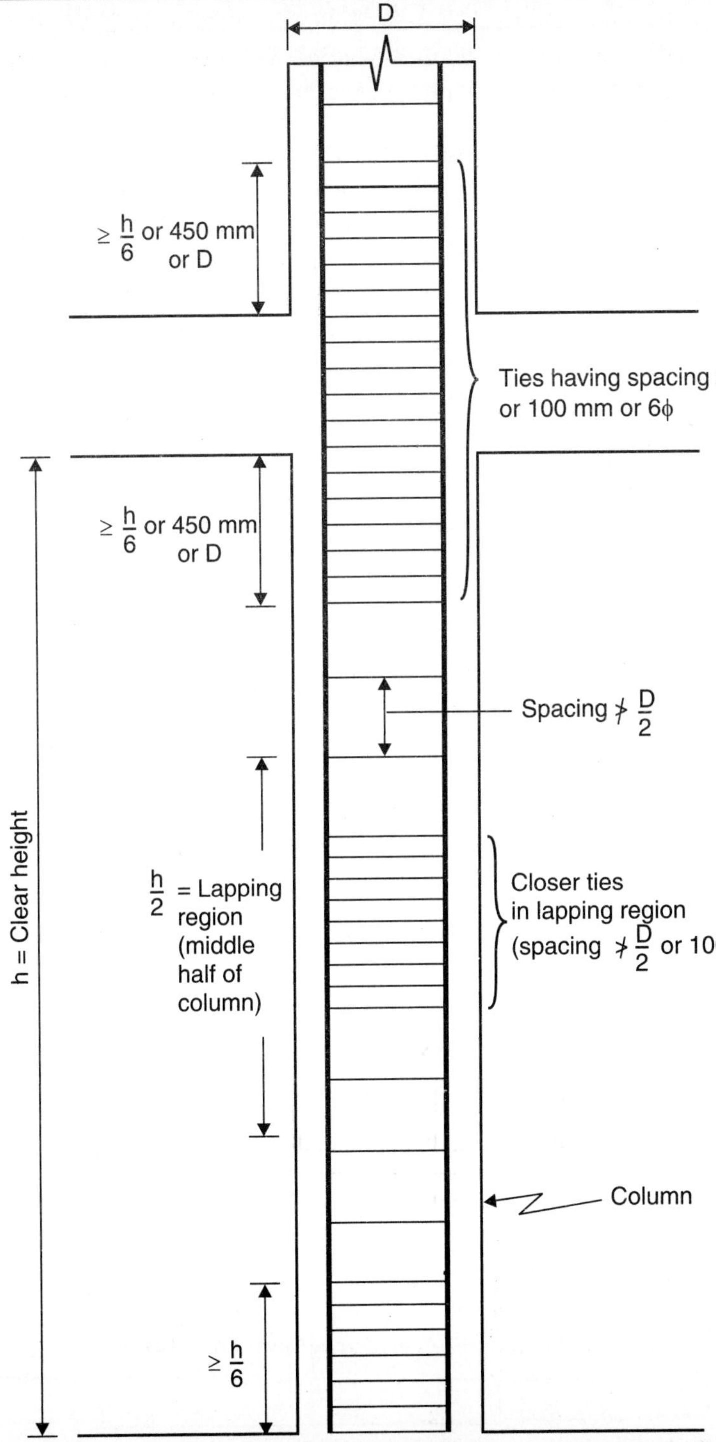

Fig. 6.6. Arrangement of ties in a column.

6.5. SPECIAL CONFINING REINFORCEMENT (CLAUSE 8.0)

- The confining reinforcement is provided over a length from the face of the joint towards midspam of beam and mid height of column on either side of the joint.
- Closely spaced ties must be provided at the ends of the column over a length not less than (*a*) larger dimension of the column, (*b*) one sixth the column height (*c*) 450 mm.
- Over the distance specified above (ends of the column), the vertical spacing of the ties in the column should not exceed D/4 (D is the smallest dimension). This spacing should not be more than 6 times the diameter of smallest longitudinal bar and not greater than 100 mm.
- When a column goes into a footing or mat, the special confining reinforcement should extend at least 300 mm into the footing or mat as shown in Fig. 6.7.

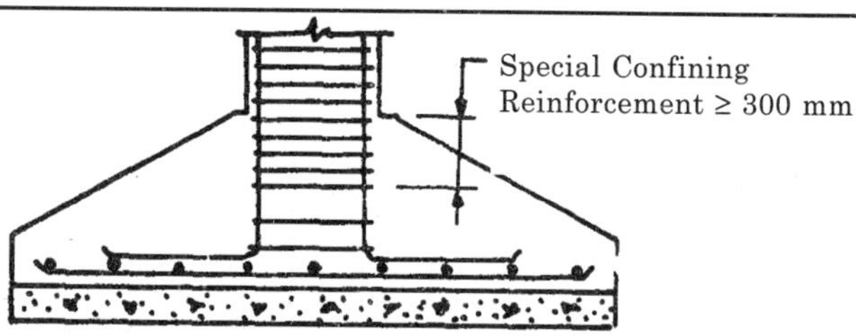

Fig. 6.7. Provision of special confining reinforcement in footing.

- Special confining reinforcement should be provided over the full length of column which has lot of stiffness variation. It may be because of presence of bracing, a mezzanine floor or a R.C.C. wall extending over a part of column height as already explained in article 5.4(*b*). Short column effect.

Figure 6.8 shows columns with varying stiffness which develop short column effect. Figure 6.9 shows special confining reinforcement required for short or captive columns.

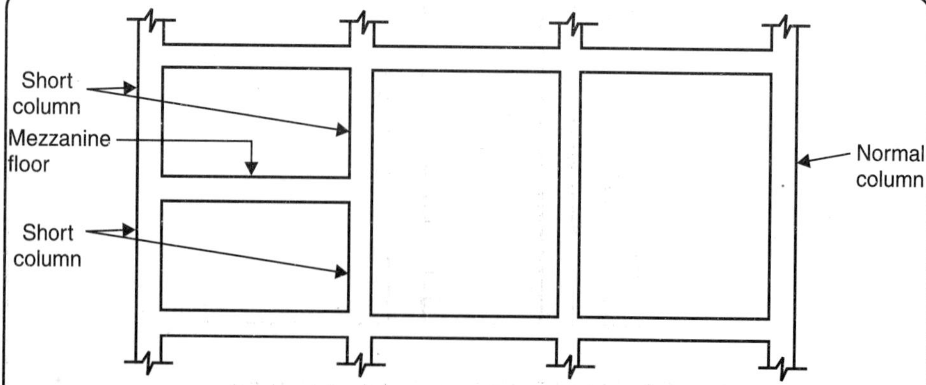

Fig. 6.8. Columns with varying stiffness.

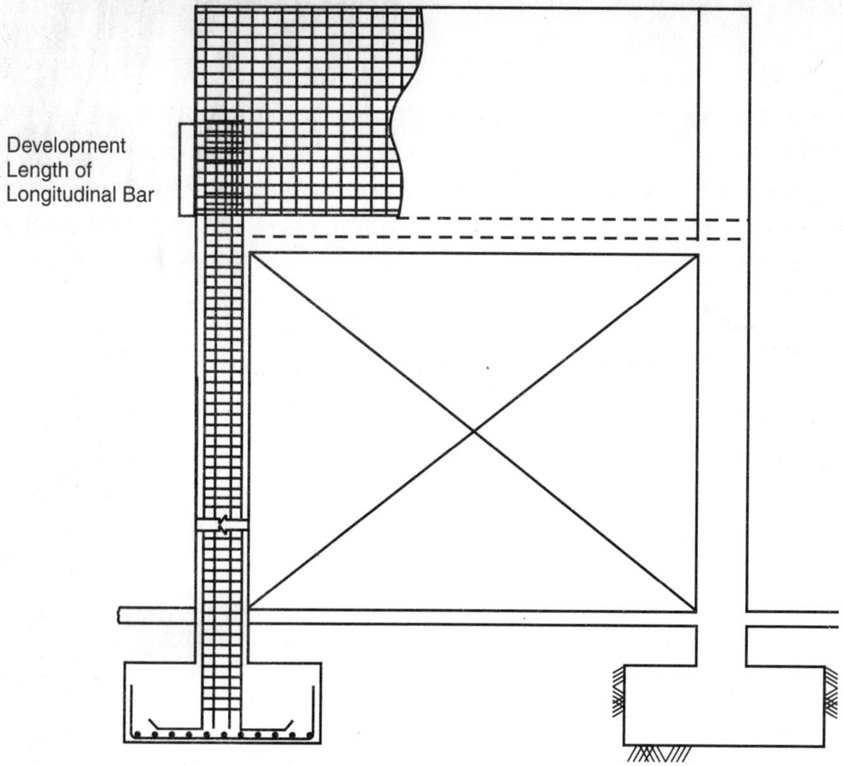

Fig. 6.9. Special confining reinforcement requirement for short or captive columns.

- The area of bar forming circular ties (spiral) to be used as special confining reinforcement shall not be less than

$$A_{sh} > 0.09 S D_k \frac{f_{ck}}{f_y}\left[\frac{A_g}{A_k} - 1.0\right] \text{ or } 0.024\, S D_k \frac{f_{ck}}{f_y}$$

where A_{sh} = Area of bar

S = Pitch of spiral

D_k = Diameter of core, measured to the outside of spiral

f_{ck} = Characteristic compressive strength of concrete

f_y = Yield stress of steel ties

A_g = Gross X-area of the column

A_k = Area of concrete core = $\frac{\pi}{4} D_k^2$

- The area of X-section, A_{sh} of the bar forming rectangular ties, as special confining reinforcement shall not be less than

$$A_{sh} > 0.18\, Sh \frac{f_{ck}}{f_y}\left[\frac{A_g}{A_k} - 1.0\right] \text{ or } 0.05\, Sh \frac{f_{ck}}{f_y}$$

where h = Longer dimension of rectangular core measured upto outside of ties and) 300 mm.

A_k = Area of core in the rectangular ties.

The following examples explains the estimation of area and spacing of spiral ties.

Example 1. Design confining spiral ties for a circular column of diameter 300 mm, grade of concrete M20 and that of steel is Fe 415.

Solution. Spacing of spiral ties S should not be more than smaller of (*a*) and (*b*)

(*a*) $\frac{1}{4} \times 300 = 75$ mm (*b*) 100 mm

Therefore $S = 75$ mm

Assuming (*i*) Clear cover = 40 mm

(*ii*) Diameter of ties = 8 mm

$$D_K = 300 - 2 \times 40 + 2 \times 8 = 236 \text{ mm}$$

$$A_g = 300 \times 300$$

$$A_k = \frac{\pi}{4} \times D_k^2 = \frac{\pi}{4} \times (236)^2$$

$$A_{sh} > 0.09 S D_k \frac{f_{ck}}{f_y}\left[\frac{A_g}{A_k} - 1.0\right]$$

$$> 0.09 \times 75 \times 236 \times \frac{20}{415}\left[\frac{300 \times 300}{\frac{\pi}{4} \times (236)^2} - 1.0\right]$$

$$A_{sh} > 81.26 \text{ mm}^2$$

X-sectional area of 12 mm diameter bar = 113 mm^2 which is greater than A_{sh}. Hence, spiral ties of 12 mm diameter bar at a spacing of 75 mm centre to centre will be adequate.

Example 2. Consider a column of 650 mm × 500 mm. The grade of concrete is M20 and grade of steel is Fe 415. Design the rectangular lateral ties for the column to be used as special confining reinforcement.

Solution. Assume : (*i*) Clear cover = 40 mm

(*ii*) Diameter of lateral ties = 10 mm

Size of core

Longer dimension = 650 – 2 × 40 + 2 × 10
= 590 mm

Shorter dimension = 500 – 2 × 40 + 2 × 10
= 440 mm

Both dimensions are greater than 300 mm hence a cross-ties is to be provided or a pair of overlapping ties as shown in Fig. 6.5.

$$h = \frac{590}{2} = 295 \text{ mm} \quad \text{and}$$

$$\text{Area of core} = A_k = 590 \times 440$$

Spacing S of links is smaller of

(a) $\frac{1}{4} \times 500 = 125$ mm (b) 100 mm

Hence, $S = 100$ mm

$$A_{sh} > 0.18 \times 100 \times 295 \times \frac{20}{415}\left[\frac{650 \times 500}{590 \times 440} - 1.0\right]$$

$$> 64.47 \text{ mm}^2$$

which is less than 78.54 mm^2 (area of 10 mm bar)

Thus, 10 mm diameter lateral ties at 100 mm c/c can be provided.

6.6. BEAM COLUMN JOINT (CLAUSE 9.0)

In RC buildings, portions of columns that are common to beams are called beam column joints. It is seen that these joints are severely damaged in earthquakes and hence these joints should be designed carefully to resist earthquake loads. The various codal provisions for Beam detailing column joints are as follows :

- The special confining reinforcement as required at the end of the column should be provided through the joint as shown in Fig. 6.10. The spacing of ties should not exceed 150 mm.

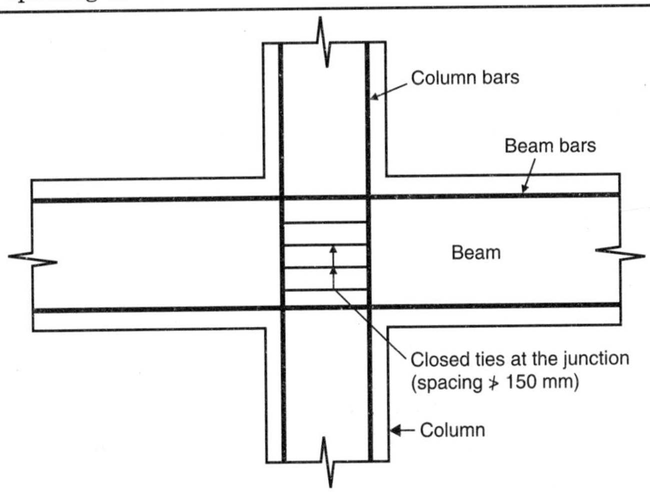

Fig. 6.10. Ties in a beam column joint.

- It is preferred to have large column size for Seismic zone III, IV and V. The column should be atleast 300 mm in each direction or 20 ds where ds is the diameter of largest bar in beam.
- In exterior joints where beams ends at columns longitudinal beam bars neeed to be anchored into the column to ensure proper gripping. The length of anchorage should be development length in tension plus ten times the bar diameter (Fig. 6.11)
- In an interior joint the beam bars (both top and bottom) should go through the joint without any cut or bent in the joint region (Fig. 6.12)

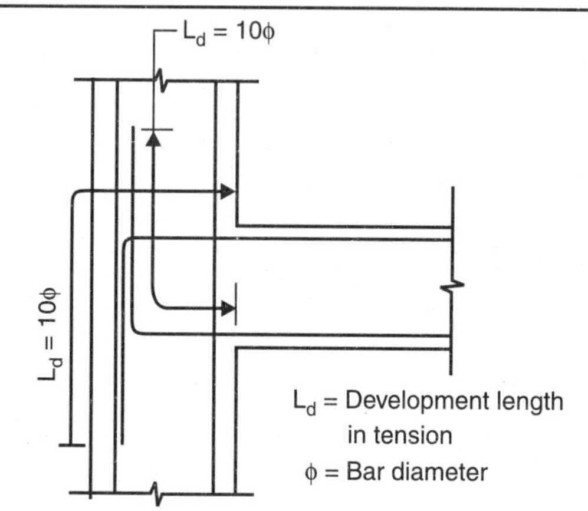

Fig. 6.11. Arrangement of beam bars in exterior joint.

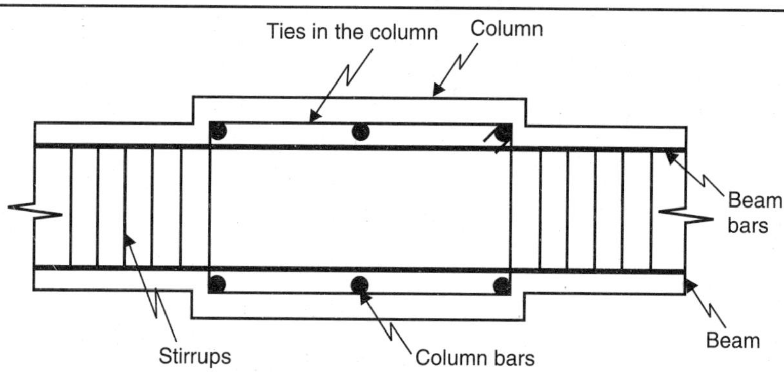

Fig. 6.12. Arrangement of beam bars in an interior joint.

6.7. SHEAR WALLS

Reinforced Concrete (RC) buildings have vertical plate like RC walls called shear walls, in addition to slabs, beams and columns. These are like wide beams which are vertically placed. There walls start at foundation level and are continuous throughout the building height. Their thickness can be as low as 150 mm or as high as 400 mm in high rise buildings. There are provided in multistoreyed buildings. Properly designed and detailed RC buildings with shear walls have shown verty good performance in past earthquakes. Figure 6.13 shows a reinforced concrete building with shear walls. The provisions for design of shear walls are given in IS 13920 : 2016 (clause 10.0). Some of them are explained below :

- The minimum ratio of length of wall to its thickness shall be 4.
- Shear walls should be provided along both length and width.
- Door or window openings may be provided in shear wall but they should be small in size and symmetrically located.
- Shear walls in buildings must be symmetrically placed.
- Steel bars should be provided in vertical and horizontal grids in shear walls.

- Minimum area of reinforcing steel should be 0.0025 times the X-sectional area along each horizontal and vertical direction. The vertical reinforcement should be distributed uniformly across the wall cross section.

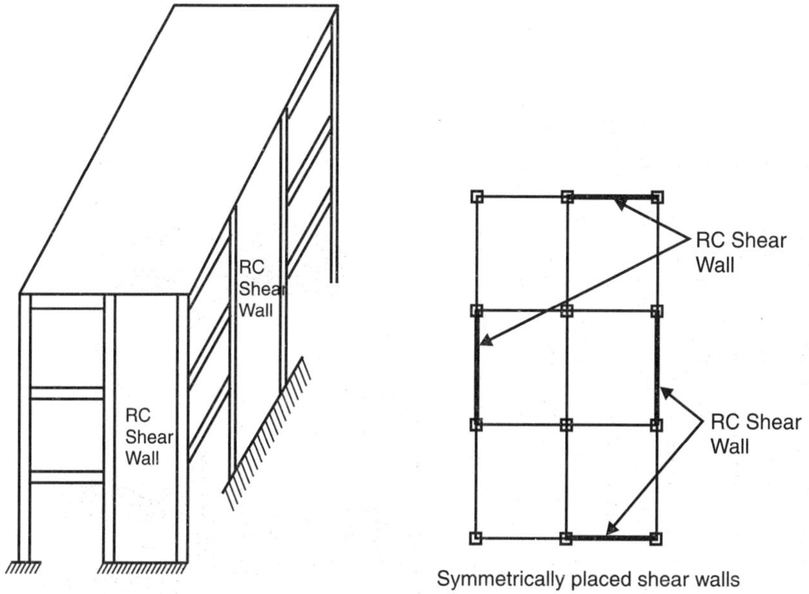

Fig. 6.13. Shear wall in RC buildings.

TERMS TO REMEMBER

1. **Flexural Members :** The members which resist bending for example beams and slabs.
2. **Development Length :** The minimum length of bar which must be embedded in concrete beyond any section to develop by bond.
3. **Stirrups :** Closed loops around the longitudinal bars which resists diagonal cracking due to shear.
4. **Columns :** The vertical compression members which are subjected to mainly axial loads.
5. **Ductile failure :** The failure which is not sudden and given sufficient warning before collapse.
6. **Anchorage length :** The minimum length provided at the end of the bar, in the form of nodes or bends, for anchoring.
7. **Splicing :** The bars which are short in length, are given a suitable lap with other bars for continuity.
8. **Shear walls :** Reinforced concrete walls which are symmetrically placed in a building to resist horizontal loads.

EXERCISES

A. Objective Type Questions :

1. IS 13920 : 2016 deals with the
   (*a*) Estimation of seismic forces
   (*b*) Ductile detailing of RC buildings
   (*c*) Strengthening of RC buildings.
   (*d*) Strengthening of masonry buildings.
2. For RC buildings which are more than 3 storeys, the minimum grade of concrete to be used is
   (*a*) M20 (*b*) M15
   (*c*) M25 (*d*) M10
3. For earthquake prone region, low strength steel is preferred over strong steel because
   (*a*) It is more brittle (*b*) It has lower breaking strain
   (*c*) It is more ductile (*d*) None of the above
4. At the ends of the beam, the amount of steel provided at bottom is
   (*a*) Equal to that at top (*b*) One fourth of that at top
   (*c*) Double of that at top (*d*) Half of that at top
5. The longitudinal bars in a beam should not be overlapped
   (*a*) At the middle of beam (*b*) At the ends
   (*c*) Anywhere in the beam (*d*) None of the above
6. The vertical stirrups should be provided in the overlapping length at
   (*a*) Wider spacing (*b*) Closer spacing
   (*c*) Same spacing as other parts (*d*) None of the above
7. To resist diagonal tension the reinforcement is provided in the form of
   (*a*) Longitudinal bars (*b*) Wire mesh
   (*c*) Stirrups (*d*) All of the above.
8. Both ends of the stirrups should be bent into a
   (*a*) 120° hook (*b*) 135° hook
   (*c*) 90° hook (*d*) 180° hook
9. The maximum spacing of stirrups should not exceed
   (*a*) Depth of beam (*b*) Twice due depth of beam
   (*c*) Half the depth of beam (*d*) None of the above
10. The ratio of X-sectional dimension of a column should not be less than
    (*a*) 0.4 (*b*) 0.5
    (*c*) 0.3 (*d*) 0.6
11. The lateral ties in a column should be provided over the entire lap length at a spacing not more than
    (*a*) 120 mm (*b*) 130 mm
    (*c*) 140 mm (*d*) 100 mm
12. The bar in a lateral ties should have an extension, after the 135° hook, not less than
    (*a*) 100 mm (*b*) 65 mm
    (*c*) 90 mm (*d*) 80 mm

13. Special confining reinforcement should be provided in the short/captive columns over
    (*a*) One fourth of the captive length
    (*b*) Half of the captive length
    (*c*) Full length of column
    (*d*) One third of captive length
14. The spacing of ties to be used as special confining reinforcement should not be greater than
    (*a*) 100 mm (*b*) 75 mm
    (*c*) 150 mm (*d*) 120 mm
15. It is preferred to have large column sizes in
    (*a*) Seismic zone III (*b*) Seismic zone IV
    (*c*) Seismic zone V (*d*) All of the above
16. The anchorage length provided in the exterior beam column joint should not be less than
    (*a*) L_d + 10 times bar diameter (*b*) L_d
    (*c*) 20 times bar diameter (*d*) L_d + 5 times bar diameter
17. Shear walls are made of
    (*a*) Brick masonry (*b*) Plain concrete
    (*c*) Reinforced concrete (*d*) Stone masonry
18. The minimum thickness of shear walls as per IS 13920 is
    (*a*) 180 mm (*b*) 400 mm
    (*c*) 200 mm (*d*) 150 mm
19. Minimum two bars should be provided in a beam through the full length at
    (*a*) Top (*b*) Bottom
    (*c*) Both (*a*) and (*b*) (*d*) None of the above
20. The number of bars to be spliced at a section should not exceed.
    (*a*) 50% (*b*) 40%
    (*c*) 25% (*d*) 30%

B. Fill in the Blanks :

1. A ________ material is the one which can undergo large elongation.
2. Ductile material gives ________ before failure.
3. Ductility can be incorporated into structures by paying attention to the ________ of reinforcement.
4. The code titled "Ductile detailing of reinforced concrete structures, subjected to seismic forces" is IS : ________.
5. Ductility ________ the extent of damage caused by earthquake.
6. As per IS 13920 : 2016, the minimum grade of concrete for RC buildings is ________.
7. Low strength steel is ________ ductile.
8. The beams and slabs are ________ members as they resist bending.
9. The ________ cracks at centre are flexural crucks.
10. The cracks at the end of the beam are diagonal tension cracks caused due to ________.

11. As per IS 13920 : 2016, the width to depth ratio of flexural members should be more than ________.
12. The lap length of the bar should not be less than the ________ length of the bar.
13. As per IS 13920 : 2016, at least ________ bars go through the full length of the beam at the top as well as the bottom of the beam
14. Transverse reinforcement in a beam are called as ________.
15. The diameter of stirrups must be at least ________ mm.
16. The maximum spacing of stirrups should not exceed ________ the depth of a beam.
17. The earthquake load is ________ in nature thus causing tension on both faces of the column.
18. The bars should be lapped only in the ________ of the column.
19. Lateral ties prevent ________ of the vertical bars.
20. The spacing of lateral ties should not exceed ________ the least lateral dimension.
21. ________ should be provided over the full length of column which has a lot of stiffness variation.
22. The spacing of special confining reinforcement at the end of the column should not exceed ________ mm.
23. Column size for seismic zone III, IV and V should be at least ________.
24. Vertical plate like RC walls are called as ________ walls.
25. Shear walls in building must be ________ placed.

C. State True/False :

1. Ductile detailing of structure can be done with the help of IS 1893 : 2002.
2. Spacing of lateral ties at the ends of the column should be less.
3. Distribution bars in slabs are always placed at bottom of main longitudinal bars.
4. Shear walls resists horizontal loads.
5. Strong steel is preferred for designing earthquake resistant structures.
6. The bars should not be overlapped in the middle of the beam.
7. Not more than 75% of the bars shall be spliced at any section.
8. Stirrup is a form of vertical shear reinforcement.
9. The lap length should be less than the development length of the bar in tension.
10. Both ends of the vertical stirrup should be bent into a 90° hook.
11. The first stirrup should be at a distance not greater than 50 mm from the face of the support.
12. Columns are the vertical compression members subjected to axial loading and bending.
13. In columns, the overlapping should be done at the ends only.
14. Ties must be provided over the entire lap length.
15. Lateral ties in columns prevent diagonal cracking.
16. As per IS 13920 : 2016, the transverse reinforcement in a circular column should be spiral or circular.
17. Beam column joints are not much damaged during earthquake.
18. Closely spaced ties must be provided at the end of the column.

19. It is desirable to have small column size for seismic zone V.
20. Shear walls are made up of brick masonry.
21. The thickness of shear wall should not be less than 150 mm.
22. Shear walls should be placed symmetrically in a building.
23. Steel is not required at the top face of the beam.
24. Ductile detailing helps in improving the seismic performance of the structure.
25. As per IS 13920 : 2016, the minimum grade of concrete for RC buildings is M15.

ANSWERS

A. Objective Type Questions :

1. (*b*)	2. (*c*)	3. (*c*)	4. (*d*)	5. (*a*)
6. (*b*)	7. (*c*)	8. (*b*)	9. (*c*)	10. (*a*)
11. (*d*)	12. (*b*)	13. (*c*)	14. (*a*)	15. (*d*)
16. (*a*)	17. (*c*)	18. (*d*)	19. (*c*)	20. (*a*)

B. Fill in the Blanks :

1. ductile	2. warning	3. detailing	4. 13920 : 2016	5. reduces
6. M20	7. more	8. flexural	9. vertical	10. Shear
11. 0.3	12. development		13. two	14. stirrups
15. 8	16. half	17. reversible		18. middle half
19. buckling	20. half	21. Special confining reenforcement		
22. 100	23. 300	24. Shear	25. symmetrically.	

C. State True/False :

1. False	2. False	3. False	4. True	5. False
6. True	7. False	8. True	9. False	10. False
11. True	12. True	13. False	14. True	15. True
16. True	17. False	18. True	19. False	20. False
21. True	22. True	23. False	24. True	25. False

SHORT ANSWER TYPE QUESTIONS

1. Explain the following terms :
   (*a*) Ductility
   (*b*) Transverse reinforcement in a beam.
   (*c*) Special confining reinforcement
   (*d*) Splicing
   (*e*) Shear walls
2. Specify the grade of concrete and the type of steel recommended by IS 13920 : 1993 for reinforced concrete building.
3. Explain terms flexural cracks and shear cracks.
4. Specify the locations where lapping of longitudinal reinforcement is prohibited in a beam.
5. What is the function of stirrups in a beam ?
6. What is the recommended diameter of stirrups as specified in IS 13920 : 2016 ?

7. Specify the zone where splicing of main bars can be done in a column.
8. Draw a diagram to show arrangement of beam bars in an exterior joint or an interior joint.
9. Why do we provide shear walls ?
10. Why is it necessary to incorporate ductility into design of RC structures ?

LONG ANSWER TYPE QUESTIONS

1. What does the seismic code IS 13920 : 1993 deal with ? Why is it necessary to incorporate ductile detailing for earthquake resistant design ?
2. With the help of a diagram show the critical section where tension may develop under earthquake loading in a frame.
3. Explain the various codal provisions for ductile detailing of longitudinal reinforcement in a beam.
4. What is the role of transverse reinforcement in a beam ? Explain various recommendations about providing stirrups in a beam as per IS 13920 : 2016.
5. What is special confining reinforcement ? Where should we provide it ?
6. With the help of a diagram explain various codal provisions for lateral ties of a column.
7. Design circular lateral ties as special confining reinforcement for a short column whose diameter is 400 mm, grade of concrete M20 and that of steel is Fe 415.
8. Design rectangular ties confining reinforcement for a column 700 mm × 500 mm. The grade of concrete is M20 and steel is Fe 415.

❑❑❑

7

CHAPTER

Introduction to IS 1893 (Part 1) : 2016

INSIDE THIS CHAPTER

7.1. INTRODUCTION

Indian Standard 1893 is based on the Seismic data collected from studies of Indian earthquakes. The first Seismic code was published in 1962 as IS 1893 "Recommendations for Earthquake Resistant Design of Structures" and later revised in 1966. As a result of additional seismic data and knowledge gained about the earthquakes, this code has been revised from time to time. The last (sixth) revision was done in 2016. In this revision, the code has been divided into the following five parts :

Part 1 : General Provision and Buildings

Part 2 : Liquid Retaining tanks—elevated and ground supported.

Part 3 : Bridges and Retaining Walls

Part 4 : Industrial Structures including stack like structures

Part 5 : Dams and Embankmenets

This chapter focuses on Part 1 which contains the general provisions applicable to all structures. This standard deals with assessment of seismic loads on various structures and earthquake resistant design of buildings. This standard also contains the seismic zone map of India as given in Fig. 1.1 of chapter 1.

As per this code, the structures are designed to resist lateral forces. These forces depend on the properties of structure and seismicity of the region. Various principles governing the design of earthquake resistant buildings and methods for estimation of seismic lateral forces are given in code IS

1893 : 2016 (Part 1). Some of the recommendations and criterion for design are discussed with example in this chapter.

7.2. SEISMIC DESIGN PHILOSOPHY

Earthquake causes motion of the ground which can be resolved in three mutually perpendicular directions. This motion causes the structure to vibrate. The vibration intensity of ground at any location depends upon the magnitude of earthquake, the depth of focus, distance from epicentre and the soil strata. The main direction of ground vibration is usually horizontal.

Severity of ground shaking at a given location during an earthquake can be minor, moderate and strong. For example, on an average about 800 earthquakes of magnitude 5.0–5.9 occur in a year and about 18 earthquakes of magnitude 7.0–7.9 (see Table) So, it is very difficult to decide whether the building is to be designed to resist a strong earthquake, the probability of whose occurrence is low or a moderate earthquake. If we design it for strong earthquake then it is too expensive. But if we design for a moderate earthquake then it may lead to major disaster, in case a strong earthquake occur. Hence, the design philosophy should lie somewhere in between these two extremes. The design philosophy is not to make buildings earthquake proof but earthquake resistant. The earthquake resistant building may get damaged severely but would not collapse during the strong earthquake and it can resist earthquakes of moderate intensities without structural damage. The IS code gives the design philosophy for earthquake resistant structures which are summarized as follows :

(*a*) Under minor earthquakes, the main members of the building that carry vertical and horizontal forces should not be damaged; but the building parts that do not carry load may get damaged.

(*b*) Under moderate earthquakes, the main members may sustain repairable damage, while the other parts of the building may be damaged such that they have to be replaced after earthquake.

(*c*) Under strong earthquakes, the main members may get severely damage but the building should not collapse.

7.3. ASSUMPTIONS

The clause 6.2 of IS code gives assumptions for the earthquake resistant design of structures which are as follows :

(*a*) Earthquake causes ground motions, which are very complex in nature. It is assumed that resonance (between the motion of structure and ground motion) will not occur as it would need time to build up.

(*b*) It is assumed that earthquake will not occur simultaneously with wind or maximum flood or maximum sea waves.

(*c*) The value of elastic modulus of materials for dynamic analysis is taken as same as for static analysis.

7.4. INCREASE IN PERMISSIBLE STRESSES IN MATERIALS

- When earthquake forces are considered along with other normal design forces, the permissible stresses in material way be increased by one third. However, for steel, the stress will be limited to yield stress or 0.2 percent proof stress.

7.5. INCREASE IN ALLOWABLE PRESSURE IN SOILS

When earthquake forces are considered the allowable bearing pressure in soils shall be increased as per Table 1 of the code depending upon type of foundation of the structure and the type of soil.

7.6. LOAD COMBINATIONS

Clause 6.3 of IS 1893 discusses various load combinations for design of structures while considering earthquake forces. For limit state design of reinforced concrete and prestressed concrete structures, the following load combinations should be used.

(1) 1.5 (DL + IL)

(2) 1.2 (DL + IL ± EL)

(3) 1.5 (DL ± EL)

(4) 0.9 DL ± 1.5 EL

where DL = Dead load

IL = Imposed load

EL = Earthquake load

+ and – sign with earthquake load is taken because earthquake load is reversible in direction.

7.7. DESIGN LATERAL FORCES

The code recommends the use of the design spectrum method for determination of lateral forces. This analysis is carried out by dynamic analysis. A simplified method may also be used which is called as equivalent lateral force procedure or static method. The main difference between the equivalent lateral force method and dynamic analysis lies in calculation of lateral forces and the distribution of lateral forces over the height of the building. The dynamic analysis of the building requires the study of dynamic behaviour of building under the dynamic loads such as earthquake loads. This study is done on the model of the structure which can be small scale model or a mathematical model. The model should represent the behaviour of the structure under assumed ground motion expected at the site. Proper modelling of the structure, material, elements and connections is required for the accurate dynamic analysis.

7.7.1. Structural Modelling

As per Clause 7.7.4.5 of IS 1893 : 2016 (Part 1). Buildings with regular, or very less irregularity may be modelled as a system of masses lumped at the floor levels with each mass having one degree of freedom that of lateral displacement. This will be clear by taking an example of a simple portal frame under lateral load analysis as shown in Fig. 7.1.

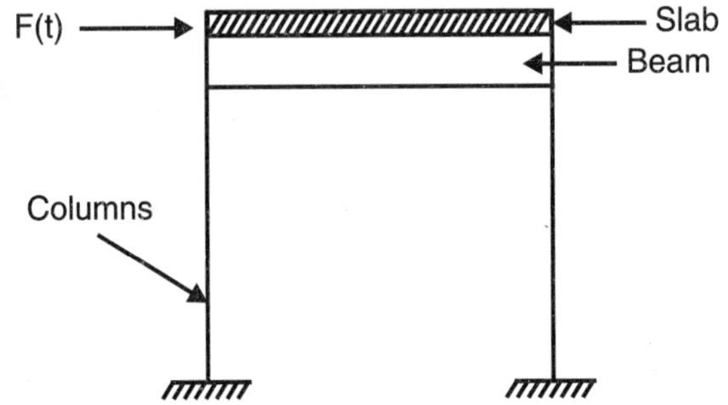

Fig. 7.1. Simple portal frame.

This system can be modelled as a lumped mass model by considering the following assumptions.

- The mass of the columns is very small in comparison to the mass of the slab. Thus, it can be assumed that the entire mass of the frame is concentrated at the slab level.
- The rigidity or stiffness of the beam and slab is very large as compared to the columns in the direction of lateral force. (The depth of column in the lateral direction is small, as compared to beam and slab). Thus, it can be assumed that the beam/slab is infinitely rigid and entire lateral deformation is due to the deformations in columns.
- The change in length of the column is negligible.
- The beam column joint is also assumed as rigid.

On the basis of above assumptions. The above system can be converted to a simple lumped mass model in which the entire mass is concentrated at the slab level and each mass is having only one degree of freedom i.e. the lateral displacement. This type of model is shown in Fig. 7.2 is very simple and most frequently used for design of multistoreyed building. This modelling reduces the calculation work for dynamic analysis of building.

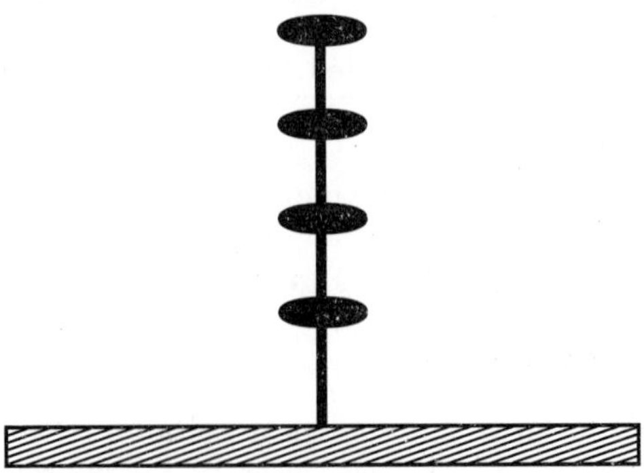

Fig. 7.2. Lumped mass model.

Other types of models are three dimension and two dimensional frame models which require a lot of computational work. They can be analyzed with the help of a computer. The buildings with irregular geometry should be modelled such that all the irregularities are depicted in the model also and generally 3-Dimensional and 2-D frame models are used for these type of buildings.

7.8. EQUIVALENT LATERAL FORCE PROCEDURE

It is the simplest method of analysis and require less calculation. In this method, each mass in a building is subjected to an equivalent lateral force. Although earthquake force is dynamic but routine design is done by assuming it to be static. The equivalent lateral force is determined according to the fundamental period of structures, given in Code IS 1893. This method is applicable for buildings with regular (uniform) distribution of mass and stiffness. In this method the design base shear shall first be calculated and then distributed along the height of the buildings as explained below :

(1) Determination of Base Shear : The total design lateral force or design base shear should be determined by the following expressions (Clause 7.5 of IS 1893).

$$V_B = A_h \,.\, W$$

where (*i*) V_B is the total design lateral force at the base of a structure.

(*ii*) *W* is the seismic weight of the buildings which is the sum of seismic weight of all floors. The seismic weight at any floor level is equal to the dead weight of floor system and appropriate amount of imposed load. The weight of floor system is dead weight of slabs, beams and half of the weight of columns and walls above and below the floor. Percentage of imposed load to be considered in Seismic weight calculation are given in Table 8 of IS 1893 are as follows

TABLE 7.1. Percentage of Imposed Load to be considered in Seismic Weight Calculation (Clause 7.3.1)

Imposed Uniformity Distributed load	Imposed load
Upto and including 3.0 kN/m^2	25%
Above 3.0 kN/m^2	50%

(*iii*) A_h = Design horizontal acceleration coefficient for a structure

$$A_h = \frac{ZIS_a}{2Rg} \quad \text{...(Clause 6.4.2)}$$

where A_h will not be taken less than Z/2

Z = Zone factor given in Table 7.2

It depends upon the Seismic zone (Seismic zone map of IS 1893 (1) : 2002) in which structure is located. Zone factors for some important towns of India are given in the Annexure I.

TABLE 7.2. Zone Factor, Z (Clause 6.4.2)

Seismic Zone	**II**	**III**	**IV**	**V**
Seismic Intensity Z	Low 0.10	Moderate 0.16	Severe 0.24	Very Severe 0.36

- I = Importance factor, it is a factor used to obtain the design seismic force depends upon the functional use of the structure. The minimum values of importance factors are given in Table 7.3.

TABLE 7.3. Importance Factors, I (Clause 6.4.2)

Sl. No. (1)	**Structure (2)**	**Importance Factor (3)**
(i)	Important service and community buildings, such as hospitals; schools; monumental structures; emergency buildings like telephone exchange, television stations, radio stations, railway stations, fire station buildings; large community halls like cinemas, assembly halls and subway stations, power stations.	1.5
(ii)	For business, community structures	1.2
(iii)	All other buildings	1.0

Notes :

1. The design engineer may choose values of importance factor I greater than those mentioned above.
2. Buildings not covered in Sl. No. (i) and (ii) above may be designed for higher value of I, depending on economy, strategy considerations like multi-storey buildings having several residential units.
3. This does not apply to temporary structures like excavations, scaffolding etc. of short duration.

- R = Response Reduction Factor. It depends on the expected Seismic damage of the structure. This factor represents the structure's ductility characteristics. The value of R increases with structure's ductility. The code IS1893 gives the value of R in Table 9 of the code for different types of buildings on the basis of experiences with past earthquakes.

- $\frac{S_a}{g}$ = Average Response Acceleration Coefficient.

It is a factor which represent the acceleration response of the structure under earthquake ground vibrations. It depends on the natural period and damping of the structure. For rock or different types of soil the values of $\frac{S_a}{g}$ are given in Fig. 7.3.

- The fundamental natural period for buildings are given in clause 7.6 of IS 1893 and they are summarized as

Earthquake Damage

$T_a = 0.075\, h^{0.75}$ For moment resisting frame building brick infill walls

$T_a = 0.080\, h^{0.75}$ For moment resisting steel frame building without in fill walls

$T_a = 0.085\, h/\sqrt{d}$ For all other buildings including moment resisting frames with brick infill walls.

where h is the height of building in m.

d is the base dimension of building at the plinth level in m, along the direction of lateral force.

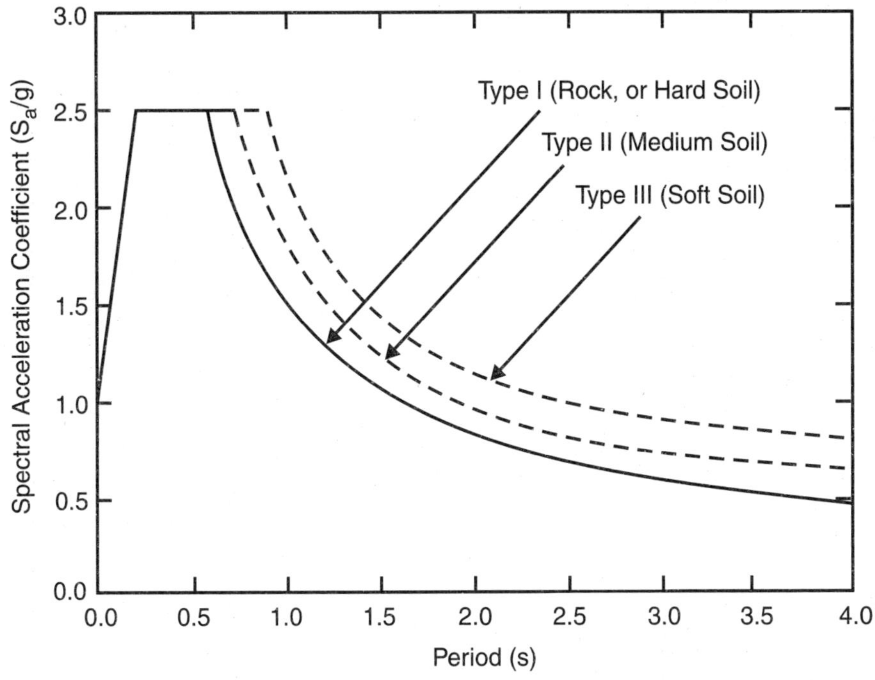

Fig. 7.3.

This shall not apply to buildings having irregular shapes. For such buildings, dynamic analysis should be done to get the natural period.

(2) Distribution of Base Shear : The base shear calculated above is to be distributed along the height of the building. The shear force at any level depends on the mass at that level and deform shape of the structure. Earthquake studies of buildings indicate that top portions suffer more damage than the lower one. Thus the top storeys should be subjected to larger forces. This is confirmed by the dynamic analysis of the buildings which are regular in shape.

The code gives the distribution of design base shear along the height of building by following expression. This gives larger forces near the top regions for the same base shear.

$$Q_i = V_B \frac{W_i h_i^2}{\sum_{j=1}^{n} W_j h_j^2}$$

where Q_i = Design lateral force at floor i

W_i = Seismic weight of floor i

h_i = Height of floor i measured from base

n = Number of storeys

For a regular (uniform building, the above equation gives an inverted parabolic distribution of forces as shown in Fig. 7.4).

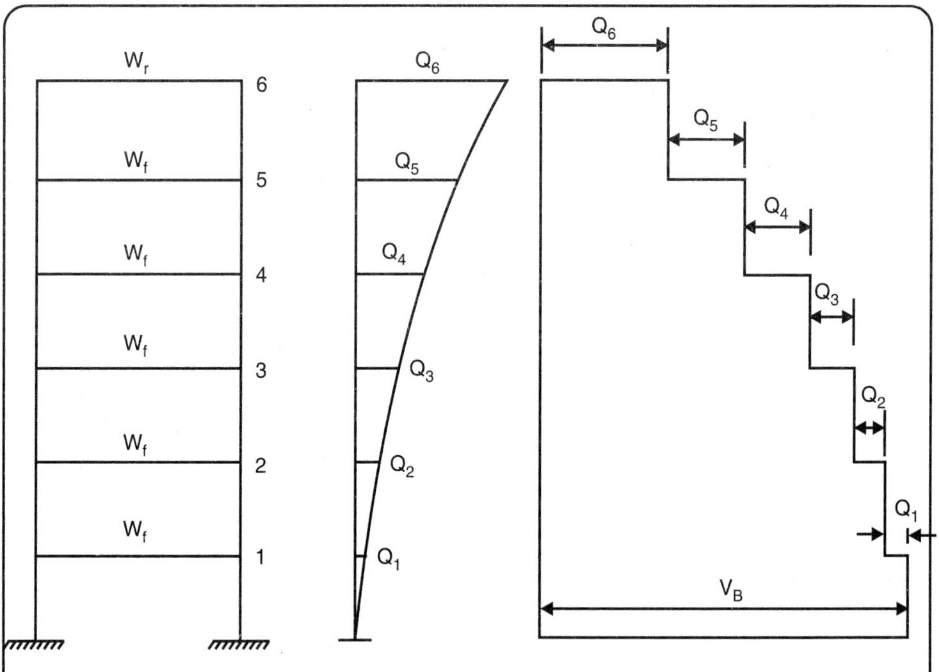

(*a*) Frame (*b*) Distribution of lateral force (*c*) Distribution of shears.

Fig. 7.4. Lateral force distribution in a uniform multistoreyed building.

All the above provisions of the code are explained below again using an example for better understanding of the concepts.

Example : One of the plane frame of a four storeyed RC building is shown in Fig. 7.5. The building is symmetric and regular. It is situated in Delhi.

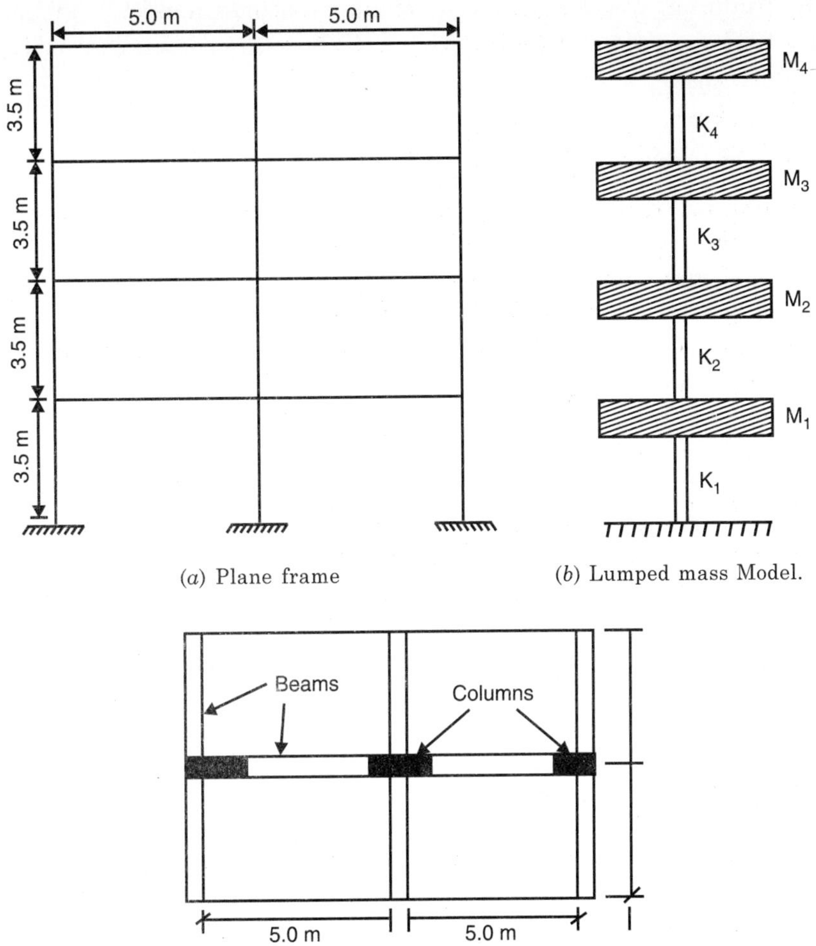

(*a*) Plane frame (*b*) Lumped mass Model.

(*c*) Plan showing the columns and beams at floor levels of the plane frame.

Fig. 7.5. Example.

Data given :

1. Floor height 3.5 m
2. Infill walls (brick) 250 mm thick longitudinal walls and 150 mm thick transverse walls
3. Live load 3.5 kN/m^2
4. Materials M20 concrete
   Fe415 steel
5. Columns 250 mm × 450 mm
6. Beams 250 × 400 mm
7. Slab 100 mm thick
8. Unit weight of RCC 25 kN/m^3

9. Building — Special RC moment resisting frame
10. Unit weight of brick masonry — 20 kN/m^3
11. Type of soil — Rocky

Solution. Step I : Calculation of lumped masses (weights) at the floor levels.

- The earthquake forces are calculated for the full dead load plus the 50 percent of live load. [Table 7.2 above 3.0 kN/m^2 – 50%]
- The live load on roof is assumed to be zero [Clause 7.3.2 of IS1893]

At roof level :

Weight of walls = Weight of walls in longitudinal direction + Weight of walls in transverse direction

$$\text{Weight of walls} = \left[0.25\times(5+5)\times\frac{3.5}{2}+0.15\times(5+5+5)\times\frac{3.5}{2}\right]\times 20$$

$$= 166.25 \text{ kN}$$

Weight of columns = Weight of 3 columns of 250 × 450 for a height of $\frac{3.5}{2}$ m.

$$= \left[3\times0.25\times0.45\times\frac{3.5}{2}\right]\times 25$$

$$= 14.765 \text{ kN}$$

Weight of beams = Weight of Longitudinal beams + Weight of transverse beams

Weight of beams = [2 × 0.25 × 0.40 × 55 + 3 × 0.25 × 0.40 × 5] × 25
= 62.5 kN

Weight of slab = 0.10 × 5 × 10 × 25
= 125 kW

Live load = 0

Weight at roof level = 166.25 + 14.765 + 62.5 + 125 + 0
= 368.51 kN

At Ist, 2nd and 3rd floor level,

Live load = 50% of live load
= 0.5 × 5 × 10 × 3.5
= 87.50 kN

Weight of walls = (0.25 × 10 × 3.5 + 0.15 × 15 × 3.5) × 20
= 332.50 kW

Weight of columns = (3 × 0.25 × 0.45 × 3.5) × 25
= 29.530 kN

Weight of beams = 62.5 kN

Weight of slab = 0.10 × 5 × 10 × 25

= 125 kN

Total weight of floor (Ist, 2nd or 3rd)

= 332.50 + 29.53 + 62.5 + 125 + 87.5

= 637.03 kN

Seismic Weight of building

= Seismic weight of all floors

= $M_1 + M_2 + M_3 + M_4$

= 637.03 + 637.03 + 637.03 + 368.51

= 2279.60 kN

Step 2 : Determination of Fundamental Natural Period.

For RC frame building without considering the brick infill panels.

$T_a = 0.075 \times h^{0.75}$ h = height of building in metre

$= 0.075 \times 14^{0.75} = 0.5423$ seconds

Step 3 : Determination of Design Base Shear

$$V_B = A_h W$$

$$A_h = \frac{Z\,I\,S_a}{2\,R\,g}$$

Z = 0.24 for zone IV (Delhi is in Zone IV)

I = 1.0

R = 5.0 (For special RC Moment resisting frame from Table 7.5)

For $T_a = 0.5423 \Rightarrow \frac{S_a}{g} = 1.842$ from Fig. 7.3

$$A_h = \frac{0.24 \times 1 \times 1.842}{2 \times 5} = 0.0442$$

$$V_B = 0.0442 \times (2279.60)$$

= 100.76 kN

Step 4 : Vertical distribution of Base Shear

$$Q_i = V_B \frac{W_i h_i^2}{\sum_{j=1}^{n} W_j h_j^2}$$

$$Q_1 = V_B \left(\frac{W_i\, h_1^2}{W_1 h_1^2 + W_2 h_2^2 + W_3 h_3^2 + W_4 h_4^2} \right)$$

$$= 100.76 \left(\frac{637.03 \times 3.5^2}{637.03 \times 3.5^2 + 637.03 \times 7^2 + 637.03\ \ 10.5^2 + 368.61\ \ 14^2} \right)$$

= 4.33 kN

Similarly $Q_2 = 17.35$ kN

$Q_3 = 39.99$ kN

$Q_4 = 40.11$ kN

Lateral force distribution at various floor levels is shown in Fig. 7.6.

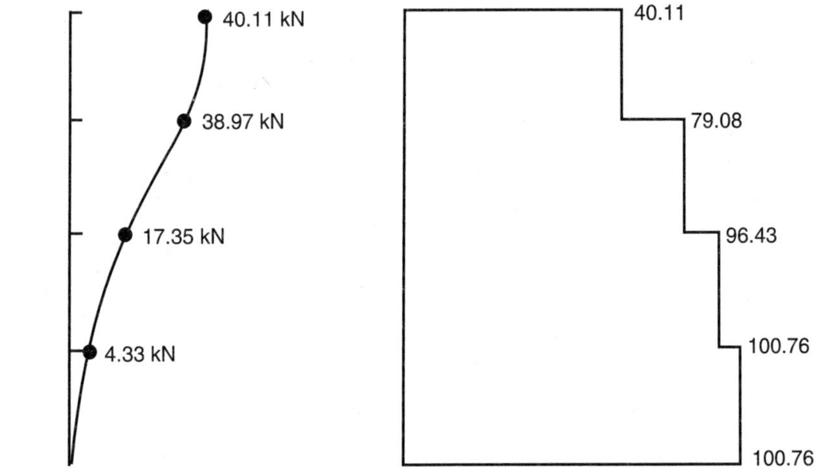

(*a*) Distribution of lateral force (*b*) Shear diagram.

Fig. 7.6. Distribution of forces at various floor levels.

7.9. DYNAMIC ANALYSIS

IS code 1893 (Part 1) : 2016 specifies the following cases when dynamic analysis should be performed to obtain the seismic design forces.

(*a*) Regular Buildings : The regular buildings having height greater than 40 m in zones IV and V and those greater than 90 m in height in zones II and III.

(*b*) Irregular Buildings : All framed buildings higher than 12 m in zones IV and V and those greater than 40 m in height in zones II and III. For irregular buildings, lesser than 40 m in height in zones II and III, dynamic analysis is not compulsory but preferred. A few buildings of this type are shown in Fig. 7.7.

Dynamic analysis is specified in clause 7.7 of code. It involves the vibration analysis of the building model having proper mass and stiffness. This analysis will give the natural periods of the building and mode shapes. **Mode shapes** are the various deflected shapes of tbe building under earthquake or dynamic loading. The natural period and the mode shapes are used for calculating the lateral forces developed due to earthquake loading. As lot of computations are involved in this method hence use of computer is a must.

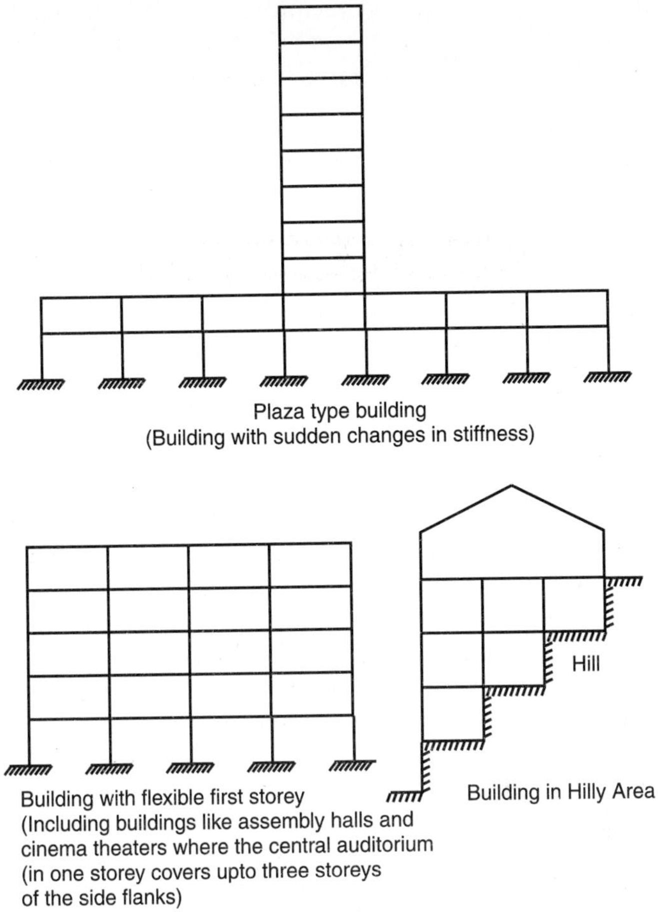

Fig. 7.7. Irregular building.

7.10. VERTICAL MOTION

The earthquake causes motion in all the three directions. It is seen that earthquake generated vertical inertia forces are not very significant for ordinary structures. Vertical acceleration should be considered in structures with large spans like bridges etc. The design acceleration for vertical motion may be taken as two third of the design horizontal acceleration (A_h) specified in art 7.6.1 clause 6.4.2 of IS 1893. Vertical designation acceleration coefficient = $\frac{2}{3} A_h$.

7.11. MISCELLANEOUS

The code IS 1893 (Part 1) : 2016 also gives provisions (clause 7.12) for projections, walls, connections between walls etc. Some of them are as follows :

1. Tower tanks, parapets chimneys and other vertical cantilever projections should be designed for five times the design horizontal

seismic coefficient (A_h). In the analysis of the building, the weight of these projections should be lumped with the roof weight.

2. All horizontal projections like balconies should be designed for five times the design vertical acceleration coefficient specified in 7.10 (that is $\frac{5 \times 2}{3} A_h = \frac{10}{3} A_h$).

TERMS TO REMEMBER

1. **Design Basic Earthquake :** It is the earthquake which can be expected to occur at least once during the design life of the structure.
2. **Design Horizontal Acceleration Coefficient :** It is the horizontal acceleration coefficient that shall be used for design of structures.
3. **Design lateral force :** It is the horizontal seismic force prescribed by IS 1893 which is used for designing structures.
4. **Design spectrum :** It is a plot of average, maximum acceleration and time period for a specified damping value for a single degree of freedom system under earthquake motion.
5. **Ductility :** Ductility of a structure or its member is the capacity to undergo large deformations without collapse.
6. **Importance factor :** If is a factor used to obtain the design seismic force depending upon the functional use of the structure.
7. **Fundamental Natural Period :** If is the first (longest) time period of undamped free vibrations of the structure.
8. **Seismic Weight :** It is the total dead load plus apropriate amount of imposed load.
9. **Design Seismic Base Shear :** It is the total design lateral force at the base of a structure.
10. **Ordinary Moment Resisting Frame :** It is a moment-resisting frame not meeting special ductile detailing.
11. **Special Moment Resisting Frame :** It is a moment resisting frame specially detailed for providing ductility as per IS : 4326 or IS : 13920.
12. **Shear Wall :** It is a wall designed to resist lateral forces in its own plane.
13. **Mode Shapes :** The deflected shapes of the building under dynamic loading.

EXERCISES

A. Objective Type Questions :

1. IS : 1893 (Part-1) : 2016 deals with
   (*a*) Ductile detailing of structures
   (*b*) Seismic strengthening of structure

(c) Calculation of seismic forces
(d) All the above

2. Earthquake is not likely to occur with
(a) Dead load (b) Wind load
(c) Line load (d) Imposed load.

3. When earthquake forces are considered in design, the permissible stresses in material are increased by
(a) 50% (b) 25%
(c) 20% (d) 33%.

4. Building with regular configuration can be modeled as
(a) Lumped mass model (b) 2-D model
(c) 3-D model (d) None of the above.

5. In a lumped mass model, the mass of the frame is assumed to be concentrated at
(a) Beam level (b) Slab level
(c) Column base (d) Centre of frame.

6. In a lumped mass model, the beam and slab are assumed to be
(a) ductile (b) flexible
(c) infinitely rigid (d) none of the above.

7. For irregular shaped building, earthquake lateral force is calculated by
(a) dynamic analysis (b) static analysis
(c) any of the above (d) none of the above.

8. The value of elastic moduls of material for dynamic analysis is
(a) Dreater than that of static analysis
(b) Less than that for static analysis
(c) Same as for static analysis
(d) none of the above.

9. The top storeys of a building are subjected to
(a) Larger forces (b) Smaller forces
(c) Some forces s other storeys (d) No forces.

10. The distribution of base shear along the height of building is

(a) $Q_i = \dfrac{V_B W_i h_i}{\sum_{j=1}^{n} W_j h_i}$ (b) $Q_i = V_B \dfrac{W_i h_i^2}{\sum_{j=1}^{n} W_j h_j^2}$

(c) $Q_i = V_B \dfrac{W_i^2 h_i}{\sum_{j=1}^{n} W_j^2 h_j}$ (d) none of the above.

11. For a regular uniform building, the distribution of lateral forces along the height of building is
(a) linear (b) hyperbola
(c) inverted parabola (d) constant.

12. The deflected shapes of the building under dynamic loading are called as
(a) deformed shapes (b) mode shapes
(c) dome shapes (d) mode shapes.

13. The design acceleration for vertical motion is equal to

(*a*) design horizontal acceleration (*b*) $\frac{2}{3} A_n$

(*c*) $\frac{1}{2} A_n$ (*d*) $\frac{1}{3} A_n$.

14. All vertical contilever projections like towers and chimneys should be designed for design horizontal seismic coefficient which is

(*a*) $5A_n$ (*b*) $2A_n$

(*c*) A_n (*d*) $3A_n$.

15. All horizontal projectional should be designed for the design vertical acceleration equal to

(*a*) A_n (*b*) $\frac{2A_n}{3}$

(*c*) $\frac{10}{3} A_n$ (*d*) $\frac{4}{3} A_n$.

B. Fill in the Blanks :

1. The first Indian seismic corde ________ was published in 1962.

2. IS : 1893, 1962 specifies ________.

3. In 2002, IS 1893 has been divided into ________ parts.

4. The seismic code IS : 1893, 2016 (Part 1) deals with ________.

5. As per IS 1893, 2016, resonance ________ occur as a result of ground motion caused by earthquakes.

6. The value of elastic modulus of material for dynamic analysis is ________for static analysis.

7. When earthquake forces are considered alongwith other design forces, the permissible stresses in materials are increased by ________.

8. ________ method is used for determination of lateral forces during earthquake.

9. The simplified method of earthquake lateral force determination is called as ________ method.

10. In the lumped mass model, the beam or slab is considered as infintely ________.

11. The entire mass of the frame is assumed to be concentrated at ________ in simple lumped mass model.

12. Three dimensional models can be analyzed only with the help of a ________.

13. The equivalent lateral force is applicable for ________buildings only.

14. The total design lateral force is called as ________.

15. The design horizontal seismic coefficient depends upon the ________ in which structure is located.

16. Delhi (zone IV) has ________ value of zone factor.

17. Bhuj comes under zone ________.

18. As per IS 1893 : 2002 India is divided into ________seismic zones.

19. The base shear is distributed along the ________ of the building as per the ________ at that level.

20. Top storeys of buildings are subjected to ________ forces than lower storeys.

21. The various possible deflected shapes of buildings under earthquake are termed as ________ shapes.
22. Dynamic analysis is preferred for ________buildings.
23. ________ factor depends upon functional use of building.
24. Zone ________ is excluded from the seismic zone map of India as per IS 1893 : 2016.
25. The seismic weight at any floor level is equal to the ________ weight of the floor system and appropriate amount of ________ load.

C. State True/False :

1. IS 1893 was first published in 1962.
2. IS 1893 : 2002 (Part 1) deals with ductile detailing of structures.
3. As per IS 1893 : 2002 earthquake is not likely to occur simultaneously with wind.
4. The permissible stresses in materials are reduced when earthquake forces are considered alongwith other normal forces.
5. The simplified method for calculation of earthquake lateral force is called as static method.
6. Importance factor depends upon the functional use of the structure.
7. Assam lies in the seismic zone IV.
8. Zone V has 0.36 as the value of zone factor.
9. The static method can also be used for irregular buildings.
10. The top stories of the buildings are subjected to smaller forces.
11. The exact method of earthquake analysis is the design spectrum method.
12. For a regular building, the distribution of base shear is inverted parabola.
13. There are five zones as per IS 1893 : 2002 seismic zoning map of India.
14. During earthquake, liquefaction generally occurs in fine saturated soil.
15. Zone I is eliminated from seismic zoning map of India as per IS 1893 : 2002.
16. Structures built in a zone will have different zone factor as per their type.
17. IS 1893 : 2002 is the fourth revision of the code.
18. For a heavy building, the design base shear is more.
19. Zone V has the least value of zone factor.
20. Earthquake causes motion in all three directions.

ANSWERS

A. Objective Type Questions :

1. (*c*)	2. (*b*)	3. (*d*)	4. (*a*)	5. (*b*)
6. (*c*)	7. (*a*)	8. (*c*)	9. (*a*)	10. (*b*)
11. (*c*)	12. (*d*)	13. (*b*)	14. (*a*)	15. (*c*)

B. Fill in the Blanks :

1. IS 1893
2. recommendations for earthquake resistant design of structures
3. five
4. General provision for buildings
5. does not
6. same
7. one third
8. Static

9. Equivalent lateral force method 10. rigid
11. slab 12. computer 13. regular 14. Design base shear
15. Seismic zone 16. 0.24 17. V
18. four 19. height, seismic weight 20. larger
21. Mode 22. Irregular 23. Importance
24. I 25. dead, imposed

C. State True/False :

1. True 2. False 3. True 4. False 5. True
6. True 7. False 8. True 9. False 10. False
11. True 12. True 13. False 14. True 15. True
16. False 17. False 18. True 19. False 20. True

SHORT ANSWER TYPE QUESTIONS

1. Name the seismic code of India. When was it first published ?
2. How do we modify the permissible stresses of materials while designing for earthquake ?
3. Name the method recommended by the seismic code for determination of lateral force for earthquake analysis.
4. Which model is best suitable for a simple partial frame for lateral load analysis ?
5. Why do we do structural modelling while doing dynamic analysis of structures.
6. Write the formula for determining base shear by static method.
7. Which type of buildings should be analyzed using response spectrum method (dynamic analysis) as specified by IS 1893 : 2002.
8. What do you understand by the term mode shape ?

LONG ANSWER TYPE QUESTIONS

1. What is the philosphy of seismic design of structures ?
2. What is a seismic code ? How do they help us in desigining structures ?
3. What are the basic assumptions for the earthquake resistance design of structures ?
4. What is structural modelling ? What is the need of modelling structures for dynamic analysis ?
5. What is a lumped mass model ? What is its use ?
6. What are the various methods of calculating design lateral for earthquake design of structures as per IS : 1893 (Part I) : 2002 ?
7. What is design base shear ? How can we calculate it ?
8. Explain the distribution of base shear along the height of the building as per IS : 1893 (Part I) : 2002.
9. What is dynamic analysis ? For which type of buildings, dynamic analysis is compulsory.
10. Determine the design base shear for the plane frame and plan of a five storeyed, symmetric RC building shown in 7.8.

    The building is situated in Jammu.

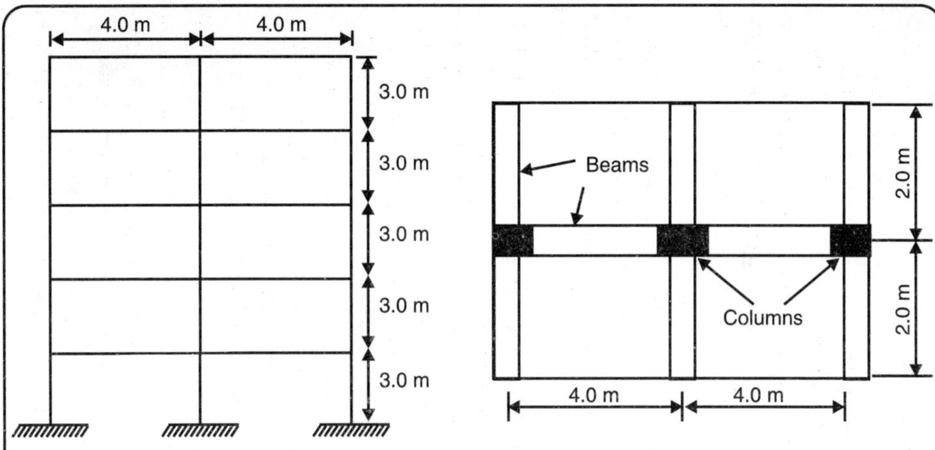

Fig. 7.8.

Data given

1.	Floor height	3.0 m
2.	Live load	3.5 KN/m^2
3.	Materials	M20 concrete Fe 415 steel
4.	Infill brick walls	250 mm thick longitudinal walls 150 mm thick transverse walls
5.	Beams	250 × 400 mm
6.	Columns	300 × 450 mm
7.	Slab	100 mm thick
8.	Unit weight of RCC	25 KN/m^3
9.	Unit weight of brick masonry	20 KN/m^3
10.	Type of soil	– Rocky
11.	Building	–Special RC moment resisting frame.

❑❑❑

8

CHAPTER

Seismic Provisions of Strengthening and Retrofitting

INSIDE THIS CHAPTER

8.1. INTRODUCTION

Till to date, earthquakes are one of the most unpredictable and devastating natural disaster which cause extensive damage to the buildings/ structures. This damage results in loss of lives and property. Thus, it is very important on the part of the civil engineers to build structures with high seismic performance. But here the question arises. "What to do with the present old/weak or earthquake damaged structures ?" It has been observed that majority of such structures/buildings may be safely reused if they are made seismically strong by using some methods/techniques (retrofitting techniques). It is a better and economical choice as compared to demolition and reconstruction. Thus there is a need to restore or strengthen the old and weak or damaged buildings so that they can sustain future earthquakes. It is one of the most important aspect of mitigation especially in earthquake prone areas which will reduce the earthquake hazards/damages. It is the most challenging task for a civil engineer in India as not much of experience and data is available in this field.

Two types of buildings need to be retrofitted.

(*i*) **Earthquake Damaged Buildings :** The buildings which are damaged or weakened by the earthquakes thus making them unfit or unsafe for future use.

(*ii*) **Weak Buildings :** Buildings which have not experienced severe earthquakes but are seismically weak and are vulnerable to earthquakes.

Various codes and standards are published by Bureau of Indian Standards to help the structural engineers in this field. This chapter explains various strengthening and retrofitting methods used for buildings as per IS 13935 : 2009. Code for "Seismic Evaluation" repair and strengthening of Masonary buildings.

8.2. NEED OF RETROFITTING

As discussed above *the need of retrofitting and strengthening is to increase the available seismic resistance of the weak/old or earthquake damaged buildings*. In addition to this retrofitting/strengthening of a building is also required to be done in the following cases :

(1) Upgradation of a Code : As the experience of the civil/structural engineers is increasing, codes/standards are also being upgraded from time to time. Thus the buildings designed by the code, which has been revised/upgraded, need to be retrofitted to fulfil the latest codal provisions.

(2) Change in use of Buildings : Whenever there are changes in the use/occupancy of a buildings, for example, public building converted to an industrial building or residential building to office building etc., there is a need to retrofit or strengthen the building to satisfy the codal provisions as per the present class of the building.

(3) Important Buildings : Important buildings such as hospitals, schools, historical monuments etc. need to be strengthened and restored from time to time to counter the effects of ageing and weathering.

(4) Retrofitting and Strengthening of a building is also needed in the case of extensions or expansion of the building, for example making more stories etc.

Replacement of damaged or unsafe buildings by reconstruction is not advisable due to following reasons :

(*i*) The cost of reconstruction is much higher than retrofitting or strengthening.

(*ii*) Historical architecture/monuments can be preserved by retrofitting.

(*iii*) Retrofitting/strengthening takes much less time as compared to reconstruction.

The relative cost of retrofitting to construction determines what is to be done. If the cost of retrofitting is less than 50% of the reconstruction cost then retrofitting is done.

8.3. TERMS ASSOCIATED WITH RETROFITTING

Retrofitting aims at increase in strength and ductility of the building/structure so that it can withstand the effects of earthquake safely. Various terms are associated with retrofitting which need to be understood carefully. These terms are explained below :

(*a*) Repair : It is defined as the process of reconstructing or renewing any damaged part of a building so that the building will have the same strength and ductility as it had before the damage.

Repair includes both structural and non-structural elements. Repair work includes filling up of cracks in masonry walls, repairing damaged beams and columns, replacing or repairing damaged utility services like water to sewage pipe lines, fixtures, gas and telephone lines etc.

(*b*) Retrofitting : It is defined as the process of increasing the seismic resistance of damaged or weak building by appropriate techniques. Retrofitting of earthquake damaged buildings is done by repairing and strengthening the damaged portions of the building and making it reusable.

Seismically weak buildings also need retrofitting so that they can sustain the effects of earthquakes safely thus minimizing loss of lives and property. It has been observed that cost of retrofitting is much less than constructing a new building.

(*c*) Restoration : This term is commonly used with buildings of historical importance *i.e.*, monuments. It is defined as the process by which a weak or damaged structure is strengthened to replicate the structure as originally built.

Structure/building may damage or some parts or elements of it become weak due to weathering, ageing or any other reason. In restoration these parts are to be repaired and strengthened so that the structure become equally strong and look exactly like it was built originally.

(*d*) Strengthening : It is the process of adding strength to any part of an existing building to provide higher strength and ductility than the original building. Strengthening can be done for earthquake damage buildings, seismically weak buildings, and also in other cases as explained in Art. 8.2.

The objectives of strengthening any structure are :

(1) Increasing the strength.

(2) Increasing the integral action of the various elements of the building *i.e.*, wall roof, foundation.

(3) Reducing sources of weakness/irregularities of the building.

(4) Increasing ductility.

8.4. SEISMIC EVALUATION OF BUILDINGS

Seismic evaluation is a process of assessing the seismic resistance of a building. This evaluation is necessary for determining or planning the retrofitting scheme or techniques required for making the structure seismically safe. The estimation of seismic resistance of a building i.e., in terms of its strength and ductility is not a simple task and require a lot of effort and studies on the part of structural engineer. There is no well defined procedure for the same. The methodologies available so far are based on the studies and examination of various parameters of the building, some of which are explained below :

(*i*) Architectural and structural drawings of the building.

(*ii*) Comprehensive examination/testing of the load bearing members of the building. It may require some non-destructive testing for assessing the strength of the material, location and amount of reinforcement and its condition (effect of rusting etc.).

(*iii*) Specification of materials, soil testing reports etc.

(*iv*) Past performance of similar type of buildings during the earthquake. The extent and type of damage in such buildings help a lot to decide about the selection of retrofitting technique/methodology.

(*v*) Visual inspection of the building which should include all types of weakness/deficiencies and damage present in the building. The weakness/deficiency include the unsymmetrical constructions, short columns, floating columns, soft stories, openings in the buildings and other points of distress like heavy floors, beam column joints etc. The damage part should include a detailed inspection of the type and extent of damage like cracking, spalling of concrete, corrosion of bars, damaged structural elements etc.

The aim of these studies is to identify the weaknesses of the buildings which help in estimating the seismic resistance of the building which is also called as Available Seismic Resistance or ASR.

The Available Seismic Resistance (ASR) of the building should be at least equal to the Minimum Seismic Resistance (MSR) specified by IS code, which is essentially required to withstand the effects of earthquake in the remaining life of the building. Thus seismic retrofitting or strengthening of a building is required to be planned and executed properly which will increase the ASR of the building to become equal to MSR.

The minimum seismic resistance of any building is estimated by design seismic coefficients given by IS 1893-2016 (Part-1) these coefficients depend upon the location of the building (zone), type of construction material of the building, importance of the building, its soil parameters, its geometry and weight etc. These parameters are discussed is detail in lesson no. 7.

8.5. METHODOLOGY FOR SEISMIC RETROFITTING

Various steps for seismic retrofitting of a building are explained below :

(1) Available seismic resistance of the building is estimated after detailed examination testing and studies as explained in Article 8.4.

(2) The seismic retrofitting and strengthening methods are recommended for the buildings based upon the detailed studies of various components of buildings and their feasibility in the present conditions.

(3) Cost analysis of the various methods/techniques is done to choose the best suitable, practical and economical option. It is suggested that the cost of retrofitting should be less than 50% of the replacement cost.

(4) A detailed and exhaustive retrofitting and strengthening procedure is planned for the building and executed in the best possible way.

(5) The upgraded/retrofitted building is analyzed as per the present code and its seismic evaluation is done to see the enhanced seismic performance.

8.6. RETROFITTING OF BUILDINGS

Various methods/techniques are used for retrofitting of buildings. These methods are broadly classified into two types :

(1) Local retrofit methods
(2) Global retrofit methods

(1) Local Retrofit Methods

These are the most commonly used methods which result into the upgradation of the individual members/elements of a building which are seismically weak. These are economical as compared to global retrofit methods. These retrofitting methods do not lead to any change in the design or geometry of the building.

For example : Repair of cracks in a wall, jacketing of beams or columns etc.

(2) Global Retrofit Method

These methods aim at increasing the seismic resistance of the structure as a whole in terms of stiffness, strength and ductility. These are costly methods and may result in change in the geometry of the building. For example, Adding new elements to the building like shear walls, infill walls, columns etc.

In the following sections, we will study suitable retrofitting techniques for traditionally built masonry buildings and R.C.C. buildings.

8.7. RETROFITTING OF TRADITIONALLY BUILT CONSTRUCTIONS

The seismic resistance of the traditionally built masonry constructions is low as compared to other type of constructions like reinforced concrete and steel. The low seismic resistance is one of the major cause of damage and casualties and is due to the following reasons :

- Masonry structures are very heavy.
- Lack of integral action : In masonry construction, the wall to wall, roof to wall and wall to foundation connections are weak and fail during the earthquake which result in collapse of the building.
- Masonry show brittle behaviour.
- Irregular/unsymmetric geometry of the building.
- Stress concentration at the openings (windows, doors etc.).
- Heavy mass.

Despite the above weaknesses, the seismic resistance of masonry buildings can be increased by suitable retrofitting techniques so that the loss of life and property can be minimized and the buildings can be safely used in future. The suitability of retrofitting techniques depend upon a number of factors some of which are listed below :

(1) Type of Structure
(2) Material of parent construction
(3) Type of masonry and its age
(4) Location and extent of damage/weakness
(5) Cost involved.

The objective of the retrofitting measures is to increase the seismic resistance of the buildings. This can be achieved by suitably selecting the retrofitting/strengthening techniques which aim at enhancing the following very important and basic requirements of an earthquake resistant masonry building :

(1) To increase the integral action i.e., all the members/elements of the building must behave as a single unit.

(2) To ensure strong and ductile connection between walls, roof and foundation of the building.

(3) To reduce and remove any type of irregularity present in the building, *i.e.*, mass, plan and elevation irregularities and try to have a symmetric, geometry if possible.

(4) To increase the strength for in plane and out of plane bending.

(5) To increase the ductility of the building.

(6) To strengthen the weaker sections/elements.

Bureau of Indian Standards has published IS 13935 : 2009 code for "Repair and Seismic strengthening of Masonry building" which includes various guidelines and methods regarding retrofitting and strengthening measures for a masonry building. Some conventional materials and methods of retrofitting of masonry constructions are explained here with their suitability conditions.

8.7.1. Repair Materials

The most common materials for repair, restoration works of various types of buildings are cement and steel. Suitable admixtures may be added to cement concrete to improve their properties such as shrinkage, bond, workability etc. Steel may be used in many forms such as bolts, rods, steel sections (beams, flats, channels, etc.), steel wire fabric etc. Wood and bamboo are used for scaffolding or providing temporary supports to the structural elements.

In addition to the above mentioned, special materials and techniques are available for repair and strengthening but these should be selected as per the nature and cost of the buildings that is to be repaired or strengthened. The availability of materials and the technical skills required for execution are also to be taken into account for deciding the repair and strengthening programmes. Some of the commonly used repair materials are as follows :

(1) Non-shrink Grouts

These grouts consists of a polymer, non-shrink cement and special sands which are mixed in definite proportions. The desired grout is to be prepared for the specific crack width. The polymer improves the adhesion of the grout with the existing masonry and also imparts higher tensile strength to the grout.

(2) Shotcrete

Shotcrete is cement mortar or cement concrete having maximum coarse aggregate size as 10 mm, conveyed through a hose and pneumatically placed under high velocity on a prepared concrete or masonry surface. The force due to pneumatic pressure compacts the shotcrete and produces a compacted, dense material. Shotcreting can be done in two ways :

(*a*) Wet mix shotcreting

(*b*) Dry mix shotcreting

In the wet mix process, all the ingredients along with water are mixed together before entering the delivery hose. In the dry mix process, the mixture of damp sand and cement is passed through the delivery hose to the nozzle where water is added. The dry mix shotcrete is used generally in the repair works of concrete elements. The bond between shotcrete and the prepared concrete surface of the damaged element is ensured with the use of suitable epoxy materials. In addition to this, suitable shear keys are provided for shear transfer between new and old concrete.

(3) Epoxy Resins

Epoxy resins are chemicals having high tensile strength which are used as binding agents. These are mixed just prior to application. They are also used for surface coating, filling cracks and holes, and also for sticking steel plates to the weak members for strengthening.

(4) Epoxy Mortar

For repairing large cracks, the epoxy resins are mixed with sand aggregate to form epoxy mortar. Epoxy mortar has higher compressive strength, higher tensile strength and a lower modulus of elasticity than concrete and hence used widely as repair material.

(5) Quick Setting Cement Mortar

This is also commonly used as repair material as the quick setting property reduces the repair time.

(6) Fibre Reinforced Plastics/Polymer (FRP)

Fibre reinforced plastic or polymer is a very versatile material for strengthening of reinforced concrete and masonry structures. It is widely used for retrofitting purpose also. FRP is 2 to 10 times stronger than steel and has high strength to weight ratio and very high corrosion resistance. FRP plates used for strengthening are glued to the wall or columns using epoxy mortars.

8.7.2. Special Techniques to restore Original Strength

(1) Grouting

Repairing of cracks is an important feature of retrofitting/strengthening. All types of cracks must be repaired as they tend to reduce the strength of masonry. Grouting is a technique, used to repair the cracked masonry walls. It consists of pressure injection of epoxy or cement mortar into the cracks. The selection of grout material depends upon the desired strength and the width of crack.

Cracks upto 1 mm width	–	epoxy injection under pressure
Crack up to 5 mm width	–	Cement mortar with epoxy
	–	reinforcement along with the (wire mesh)

The stepwise procedure for grouting is as follows :

(*i*) Clean the outer surface of the wall and remove the plaster from area near the cracks. Clean it thoroughly with air and water jets.

(*ii*) Plastic injection ports are placed along the cracks on both sides of the member at suitable spacing.

(*iii*) Port should be secured in place with some sealant. After the sealant has set, a pressure injection of epoxy resin or cement grout is applied at the bottom of the crack, if it is a vertical crack and at one end of crack, if it is a horizontal crack.

(*iv*) The process of injecting grout is done till the epoxy or cement grout penetrates into the crack fully and is seen flowing from opposite side of the member.

Note : The smaller the crack, higher is the pressure and closer ports.

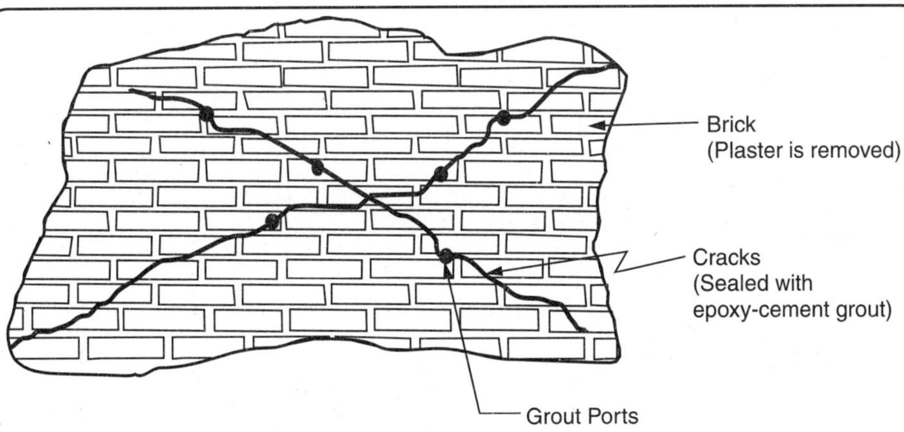

Fig. 8.1. Grout or epoxy injection in cracks

(2) Guniting

For cracks wider than 5 mm, or for areas where the masonry or concrete has crushed, the process of repair is slightly different, and is called as guniting.

Guniting is application of expansive cement mortar or quick setting mortar or epoxy cement mortar, pneumatically on the surface of the masonry in the form of slab. If may be done on both the sides of the wall, if necessary. The thickness of granite layer is 8-10 cm. The stepwise procedure for guniting is as follows :

(1) Remove all loose material and prepare the surface by making it rough and wet.

(2) Apply Gunite (expensive cement mortar or quick setting cement etc.) under pressure on the prepared surface.

(3) In the case of crushed walls, which need to be strengthened, reinforcement in the form of wire mesh (see Fig. 8.2) is placed on the damaged/cracked surface and nailed/bolted properly to the wall, then it is covered by mortar to give strength and protection to the reinforcement.

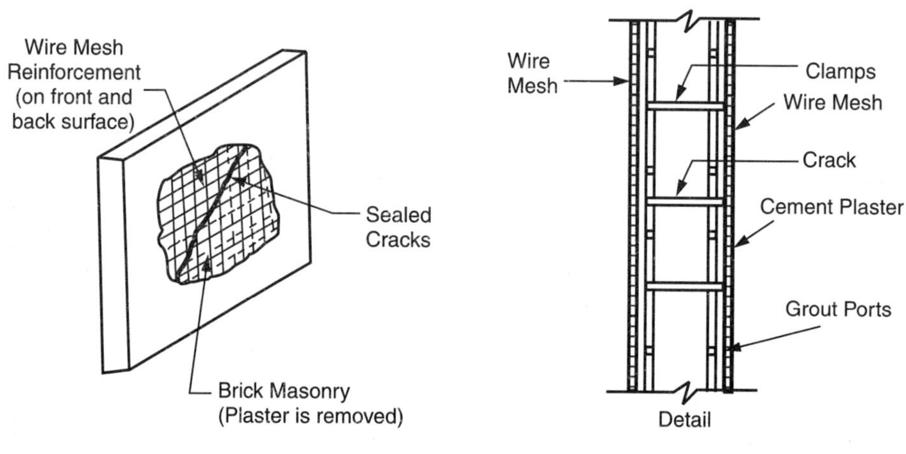

Fig. 8.2. Wire mesh reinforcement with cement mortar in cracks.

(3) Insertion of through/Bond Stones (Stone Masonry)

Stone masonry houses have very thick walls which tend to bulge and separate into layers. For such walls, "Through Stones" of length equal to thickness of the wall are inserted at regular intervals into the wall along the vertical as well as horizontal direction. These stones are fixed with cement sand mortar. If full length through stones are not available then "S" shaped steel bars or hooks are inserted into the wall as shown in Fig. 8.4.

In the care of bulged wall the affected portion of the wall is removed, reconstructed and voids are filled with cement grout.

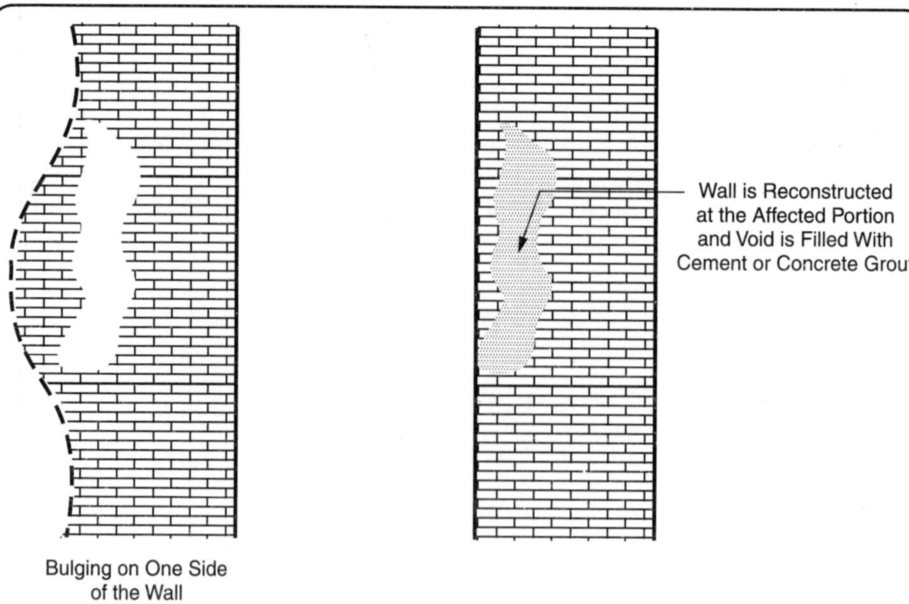

Fig. 8.3. Crack Repair in Stone Masonry.

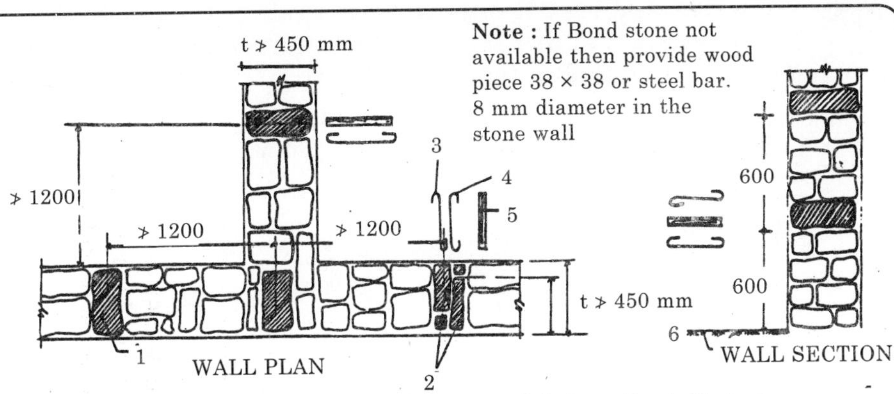

1. Bond stone, 2. Pair of overlapping stones, 3-S shape tie, 4. Hooked link, 5. Wood plank, 6. Floor level.

Fig. 8.4. Use of through stones.

(4) Shotcrete

It is a very important technique to strengthen the load bearing masonry walls which are weak or damaged. Shotcrete is cement mortar or cement concrete mortar (Coarse aggregate size ≯ 10 mm) conveyed through a hose pipe and pneumatically applied to a prepared concrete or masonry surface under high velocity. The force of application is such that it compacts the shortcrete material and produce a dense homogeneous surface. The bond between the prepared surface of the damaged wall/elements with shotcrete layer is to be ensured by applying suitable epoxy material. The stepwise procedure for shotcrete is as follows :

1. Clean the surface and remove all loose material.
2. Wet the surface properly and make it rough for bonding. If required, apply suitable epoxy material for adhesion.
3. Apply shotcrete very carefully so that it forms a homogenous layer and the appearance of building is not altered.
4. Proper curing time is given to avoid shrinkage and cracking.

In the case of damaged masonry wall, which needs to be strengthened, Reinforced shotcrete (reinforcement is also put before shotcreting) is used which is very effective and commonly used now-a-days.

(5) Splint and Bandage System

It is a very common method of strengthening the masonry buildings. It is done by providing vertical and horizontal reinforcement (wire mesh) on both outer and inner surfaces at critical sections of the building. The horizontal bands are called as bandages and the vertical steel is called as splint. The welded mesh should be nailed properly to the masonry and then it is covered with a proper thickness of micro-concrete. The advantages of this system are following :

(1) The reinforcement provides additional strength to the masonry.

(2) Ductility is increased.

(3) The splint and bandage system also binds the masonry elements (walls) together so that they act as one unit *i.e.* structural integrity and box action is achieved.

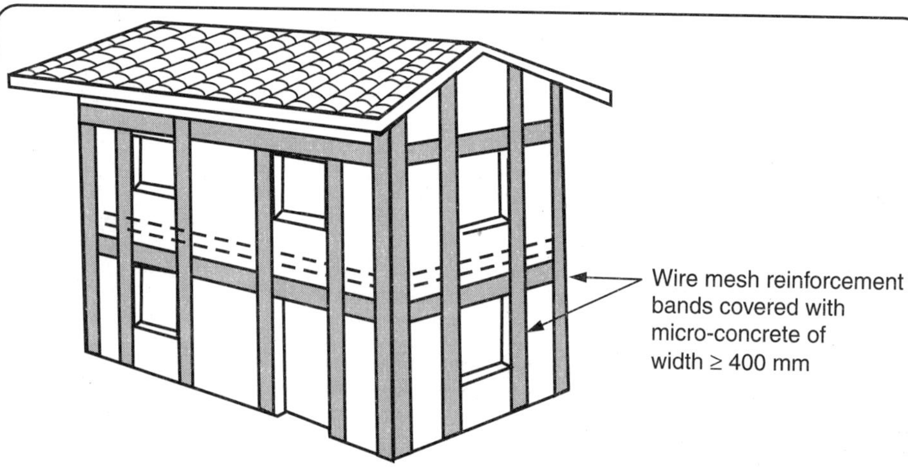

Fig. 8.5. Splint and bandage technique.

(6) Confining the Masonry

In order to increase the structural integrity, reinforcement concrete elements, (for example, columns) may be inserted, around the wall panels or in the middle of the wall. This will not only make the masonry confined but also increases the ductility and the strength.

Confinement with steel sections is also done for retrofitting the brick masonry because it is speedy and fabricated steel sections are easily available in many sizes.

Fig. 8.6. Confining masonry with reinforced concrete columns.

(7) Inserting New Walls

Unsymmetrical buildings are more prone to damage during earthquakes. Any type of discontinuity or unsymmetry obstracts the load transfer and hence should be avoided. In the case of such existing buildings, following techniques may be used for making them symmetrical.

(1) By separating the parts of the building so that the separated parts are symmetrical.

(2) Inserting new walls, may be of masonry or RCC and connecting (bonding) it properly to the old walls.

The connection between the new and old wall is to be ensured and can be achieved in may ways as explained in Fig. 5, 6, and 7 of IS13935 : 1993. In some cases, a number of keys are made in the old wall, steel is inserted in them and concrete is poured in connection between the old and new wall can also be achieved by inserting steel bars in a drilled hole in the old wall and filling it with cement grout.

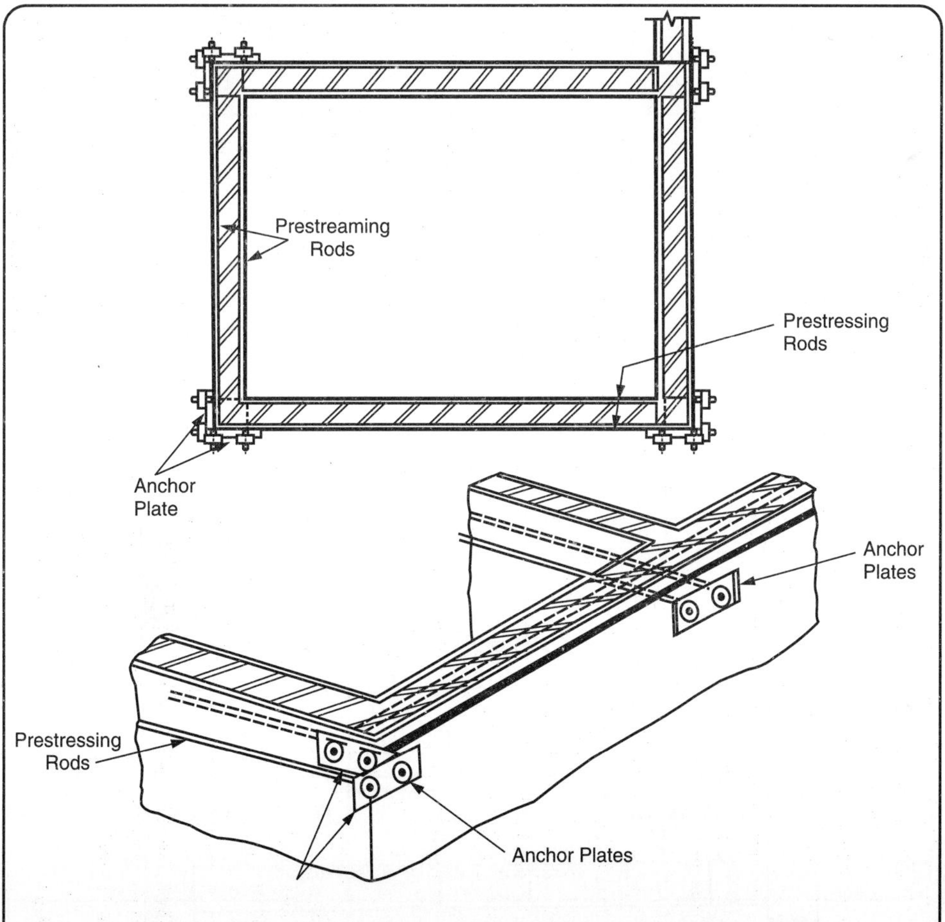

Fig. 8.7. Strengthening of weak wall by prestressing.

(8) Prestressing

Prestressing is a technique by which internal stresses of suitable magnitude are induced in the structure so the stresses from the external loads are counteracted to desired degree prestressing is used for strengthening and retrofitting of seismically weak and damaged walls.

Prestressing not only increase the strength of the wall but also helps in achieving integral behaviour of the walls (box-action).

8.8. RETROFITTING OF R.C.C. BUILDING

(1) Local Retrofitting Methods

Various types of local retrofitting methods are used for enhancing the strength of the individual elements (beam, column etc.) of R.C.C. building. Some of there are explained below :

(i) Jacketing : It is the most commonly used technique for strengthening columns and beams in a weak, old or damaged R.C.C. building. Jacketing results in increased moment resisting capacity of the member due to additional steel reinforcement. It also enhances ductility. Fig. 8.8 shows various techniques of jacketing beams and columns to strengthen them. (Fig. 8.8 and Fig. 8.9)

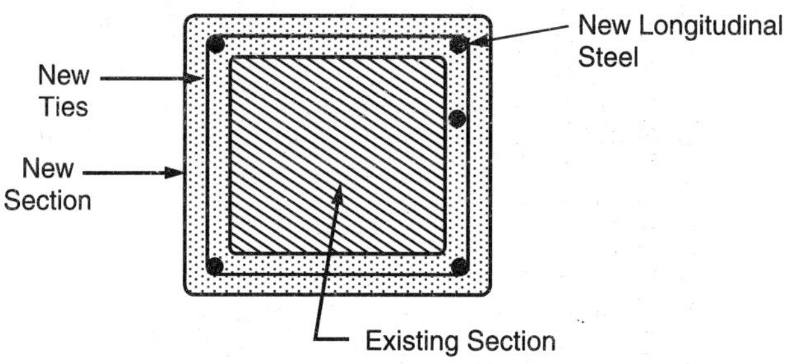

Fig. 8.8. Jacketing a column.

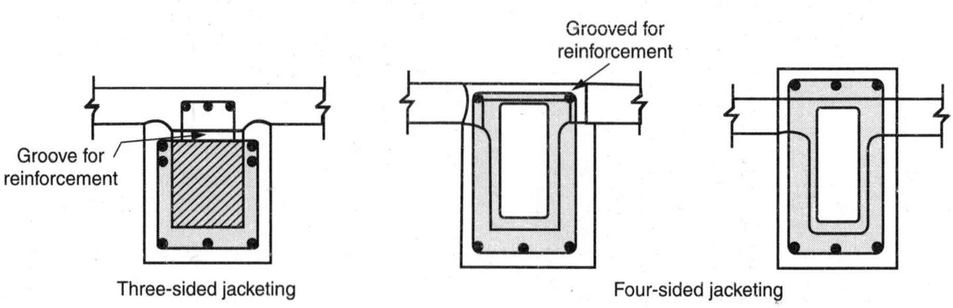

Fig. 8.9. Jacketing a beam.

- For strengthening columns, an additional area of longitudinal steel and lateral reinforcement is provided all around the existed column and concrete is poured in to cover it properly. It results

(a) Jacketing of Beam and Column.

(b) RC Shear Wall

in increase in strength and ductility thus enhancing the seismic resistance.

- Jacketing of RCC beam can also be done by providing extra reinforcement (main as well as shear) and covering it by proper thickness of concrete cover. For holding the stirrups, holes may also have to be drilled through the slab. This technique results in increase in moment of resistance and ductility of the beam section.
- Seismically weak columns are also strengthened by jacketing it with steel plates. In this method steel plates are glued to the existing column by epoxy material and by anchor bolts. The gap between the jacket and concrete is filled with cement grout.

(*ii*) **Use of Fibre Reinforced Plastic/Polymer :** Use of fibre reinforced plastic sheets is very popular now-a-days. There are used for strengthening beams, columns and slabs.

Carbon fibres or polymer fibres used in FRP's are light weight, flexible and strong, which result in increased strength of the member.

- For strengthening columns, FRP sheets are glued on to the concrete all around using epoxy resin.
- For beams, the FRP sheets are used as follows :
  (*a*) For strengthening of the tension zone, (by gluing FRP sheet to the tension face)
  (*b*) For strengthening against shear (By gluing FRP sheet on the vertical sides of the beam near the supports).

Note :

1. Before sticking FRP sheets the cracks should be repaired with epoxy resin.
2. Cement plaster or shotcrete can be done over FRP sheets to protects them from weathering.

(*iii*) **Replacing the Damaged/Worn Out Steel :** In the case of very old or damaged beams and columns, sometimes the reinforcement is worn out buckled or damaged such that it does not serve its purpose. In these cases, the concrete cover is removed from the affected portion and worn out or buckled steel is removed. New steel reinforcement is placed and welded properly to the old steel and cover is provided by high strength cement grout.

(*iv*) **Prestressing :** Prestressing is a method by which the load carrying capacity of the beams can be increased. In this method internal stresses of suitable magnitude are induced in the beam so that the stresses resulting from external loads are counteracted to desired degree. Prestressing also results in better shear resistance of the beam.

(2) Global Retrofit Methods

These methods result in the overall increase in the seismic resistance of the structure. These are costly methods as compared to local methods and generally cause change in the design/geometry of the structure. Most common methods under this category are discussed below :

(i) Adding Shear Walls/Infillwall/Bracing

- Shear walls are used for increasing the lateral strength of RC buildings. Thus addition of new shear walls is the best and simple method for improving the seismic performance of a building the new shear wall should be placed in a symmetrical manner.

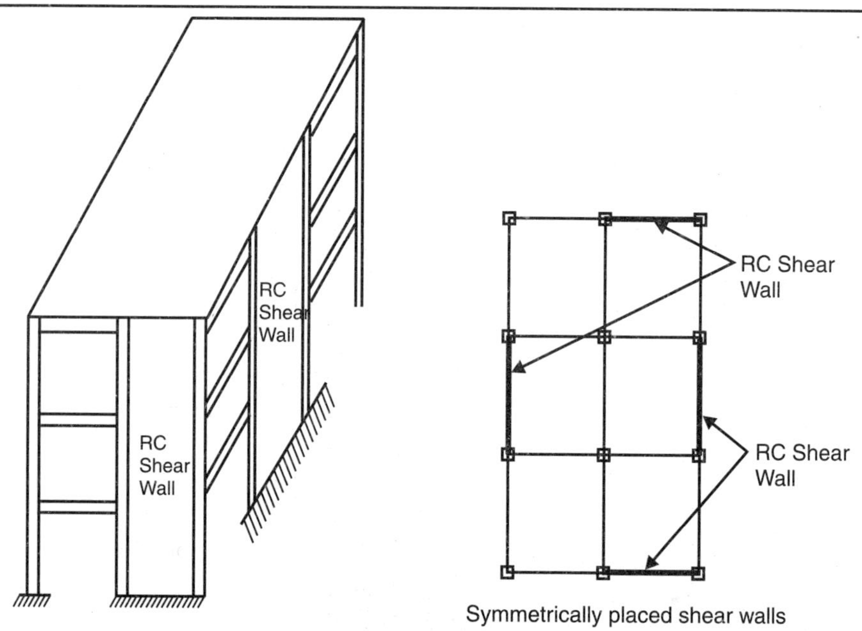

Fig. 8.10. Shear wall.

- Infill walls are also used with proper anchorage to the frame as a retrofitting technique for improving strength of the building.

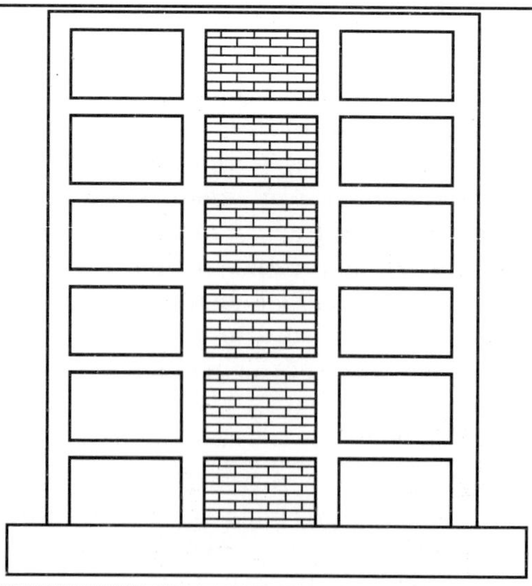

Fig. 8.11. Use of infill walls.

- Steel bracings may be used for increasing strength and stiffness of the building.

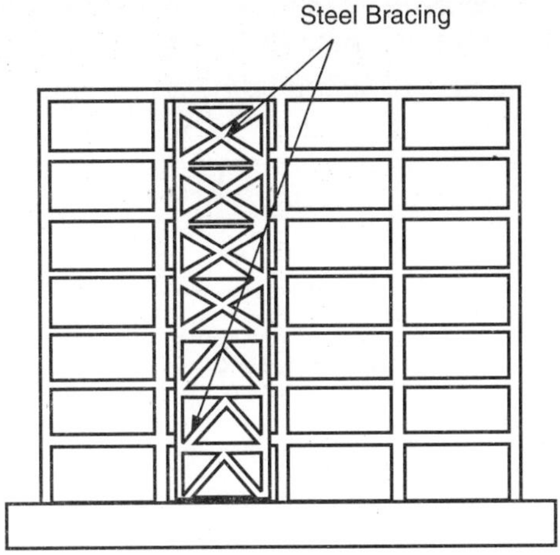

Fig. 8.12. Use of steel bracing.

(ii) Non-Conventional Methods

(*a*) **Seismic Base Isolation :** This is a non-conventional and costly method for increasing the seismic resistance of the R.C.C. building. In this method, some type of support (mainly rubber bearings) that isolates the building from the ground, is introduced between the building and foundation.

During earthquake when ground shakes, the bearings will move freely, thus no force is transferred to the building.

(*b*) **By Seismic Dampers :** It is also a non-conventional and costly method for controlling seismic damage in buildings and improving their seismic performance. In this method, seismic dampers are installed in the building at appropriate places which act as shock absorbers and thus reducing the effects of earthquakes.

Note : Non-conventional methods are explained in Annexure-II of this book.

TERMS TO REMEMBER

1. **Repair :** It is defined as the process of reconstructing or renewing any damaged part of the building so that the building will have the same strength as it had before the damage.
2. **Retrofitting :** It is defined as the process of increasing seismic resistance of a weak or damaged building by suitable techniques.
3. **Restoration :** It is defined as the process of strengthening a weak or damaged structure to replicate the structure as it was originally built.
4. **Grouting :** It is a technique used to repair the cracked masonry walls by pressure injection of epoxy or cement mortar into the cracks.
5. **Guniting :** Guniting is application of expansive cement mortar or epoxy cement mortar, pneumatically on the surface of the masonry walls in the form of slab

6. **Shotcrete :** It is cement mortar or cement concrete mortar conveyed through a hose pipe pneumatically under high velocity, applied to a concrete or masonry surface for strengthening.
7. **Prestressing :** It is a technique by which internal stresses of suitable magnitude are induced in the structure to counter the external loads to a desired degree.

EXERCISES

A. Objective Type Questions :

1. IS code for Repair and Seismic Strengthening of masonry buildings is :
   (*a*) IS1893 (*b*) IS 456
   (*c*) IS 13935 (*d*) IS 4326
2. Process of adding strength to a weak or damaged building is known as:
   (*a*) Repair (*b*) Retrofitting
   (*c*) Restoration (*d*) Reconstruction
3. Process of removing the superficial defects is called as :
   (*a*) Repair (*b*) Retrofitting
   (c) Restoration (*d*) Reconstruction
4. ASR stands for:
   (*a*) Automatic Seismic Resistance
   (*b*) Available Seismic Resistance
   (*c*) Accelerated seismic Resistance
   (*d*) None of these
5. MSR stands for
   (*a*) Mean seismic resistance (*b*) Maximum seismic resistance
   (*c*) Minimum seismic resistance
   (*d*) None of these
6. For a seismically strong structure ASR should be:
   (*a*) Equal to MSR (*b*) Less than MSR
   (*c*) Greater than MSR (*d*) None of these
7. Grouting is used for:
   (*a*) Strengthening beams (*b*) Strengthening columns
   (*c*) Repairing cracks (*d*) Strengthening floors
8. RCC beams and columns can be strengthened by :
   (*a*) Grouting (*b*) Guniting
   (*c*) Splint and bandage (*d*) Jacketing
9. Splint and Bandage technique is used for strengthening;
   (*a*) RCC buildings (*b*) Masonry buildings
   (*c*) Steel buildings (*d*) None of these
10. RCC building can be strengthened by
    (*a*) Adding shear wall (*b*) Adding infill wall
    (*c*) Adding steel bracings (*d*) All of these
11. Vertical bands used to strengthen masonry wall called as
    (*a*) Splints (*b*) Bandages
    (*c*) Columns (*d*) Verticals
12. Splint and bandage techniques are used for
    (*a*) Strength (*b*) Ductility
    (*c*) Integral action (*d*) All of these

13. Through stones are used to strengthen :
    (*a*) RCC buildings (*b*) Brick masonry
    (*c*) Stone masonry (*d*) Steel buildings
14. The technique in which super structure and foundation are separated by bearings or flexible layer is known as :
    (*a*) Baes isolation (*b*) Super structure isolation
    (*c*) Foundation isolation (*d*) None of the above
15. Indian Standard for Earthquake Resistant Design of Structures is:
    (*a*) I892 (*b*) 1893
    (*c*) 1894 (*d*) 1895

B. Fill in the Blanks :

1. IS Code ________ is the code for repair and seismic strengthening of masonry buildings.
2. ________ is the process of adding strength to the weak and earthquake damaged structures.
3. Small sized cracks (less than 5 mm width) can be repaired by ________ technique .
4. Retrofitting measures help in increasing ________ and ________ of the structure.
5. For earthquake resistant construction, available seismic resistance should be ________ than minimum seismic resistance.
6. The cost of retrofitting is generally ________ than the cost of reconstruction.
7. ________ is the process of increasing strength of RCC beams and columns.
8. ________ is the binding material, commonly used for repair works.
9. In ________ method, the structure and foundation are separated by bearing or flexible layer.
10. ________ retrofit method aims at increasing the seismic strength of the individual member.
11. ________ retrofit method aims at increasing the seismic strength of the whole structure.
12. Addition of ________ and ________ result in overall increase in seismic strength of buildings.
13. Fibre Reinforced Plastics (FRP) have ________ tensile strength and ________ weight.
14. Unsymmetrical buildings have ________ seismic resistance.
15. Reinforced concrete beams can be strengthened by applying ________.

C. State True/False :

1. Retrofitting is used for increasing the seismic strength of weak, old and damaged structure.
2. The cost of retrofitting is much more than the cost of construction.
3. Restoration results in regaining the lost strength.
4. Foundation should not be built on loose soils.
5. IS 1893:2002 is the code for repair and retrofitting of structures.
6. Vertical and horizontal steel bands are used for strengthening masonry structures.
7. Shotcrete is used for repairing cracks.
8. Jacketing is used for strengthening R.C.C. beams and columns.
9. Masonry has low weight and high tensile strength.

10. Bands increase the integral action of masonry walls.
11. Prestressing is used for strengthening R.C.C. beams.
12. Liquefaction occurs in clayey soil.
13. Asymmetrical buildings show high seismic resistance.
14. Addition of shear wall increases the seismic resistance of R.C.C. buildings.
15. Global Retrofitting techniques improves strength, ductility and integral action of the building.

ANSWERS

A. Objective Type Questions :

1. (*a*)	2. (*b*)	3. (*a*)	4. (*b*)	5. (*c*)
6. (*b*)	7. (*c*)	8. (*d*)	9. (*b*)	10. (*d*)
11. (*a*)	12. (*d*)	13. (*c*)	14. (*a*)	15. (*b*)

B. Fill in the Blanks :

1. 13935:1993	2. Retrofitting	3. Grouting
4. Strength, ductility	5. Greater	6. Less
7. Jacketing	8. Epoxy	9. Base Isolation
10. Local	11. Global	
12. Shear walls, Steel bracings		13. High, low
14. Low	15. Prestress	

C. State True/False :

1. True	2. False	3. True	4. True	5. False
6. True	7. False	8. True	9. False	10. True
11. True	12. False	13. False	14. True	15. True

SHORT ANSWER TYPE QUESTIONS

1. Define the following terms :
   (*a*) Repair (*b*) Restoration
   (*c*) Retrofitting (*d*) Strengthening
2. What is the need of retrofitting ?
3. Differentiate between local and global methods of retrofitting.
4. Explain the method of repair of cracks in masonry buildings.
5. What is jacketing and how it is done? Explain with the help of diagrams.

LONG ANSWER TYPE QUESTIONS

1. What is retrofitting ? why is it done ?
2. Explain the method of seismic evaluation of a building.
3. Explain any three retrofitting techniques used for masonry buildings.
4. Explain the techniques used for strengthening RCC beams and Columns.
5. Explain the non-conventional methods of retrofitting used for RCC buildings.

❑❑❑

9

CHAPTER

Disaster Management

INSIDE THIS CHAPTER

9.1. INTRODUCTION

A disaster is a phenomenon which puts humans and property in a dangerous and disadvantageous situation. The disasters can be natural or man-made.

Natural Disasters

- **Wind related :** Cyclone, storm, hurricane, tornado etc.
- **Water related :** Floods, cloud burst, drought
- **Earth related :** Earthquake, tsunamis, avalanches, landslides, Volcanoes.

Man-Made Disasters

- War/battle/riots
- Accidents of trains/vehicles/aeroplanes
- Industrial accidents
- Fire and forest fires
- Nuclear explosions/accidents
- Deforestation/soil erosion/air/water pollution.

These disasters are regular phenomenon in the world. Whatever be the type of disaster, the result is loss of lives and damage to the properties. Thus, it is necessary to reduce these effects by planning before hand. Disaster management is the discipline dealing with and avoiding risks caused due to hazards/disasters. It involves preparing, supporting and rebuilding society when natural or man-made disasters occur.

9.2. DISASTER MANAGEMENT

Some of the disasters can be predicted while some cannot. Earthquake prediction is not possible till to date. Some one has rightly said disasters will always come and we will have to live with them. Every year mainly lives are lost due to earthquake. Some of them can be saved by following disaster management principles. The most important activity after earthquake occurrence is Rescue. A properly managed Rescue operation can result in saving of lives. Thus, this chapter focuses on the essentials of rescue operation like rescue workers, rescue plan, safety aspects, casualty management etc. The measures to reduce the adverse effects of disasters is disaster management. It consists of four phases as shown in Fig. 9.1.

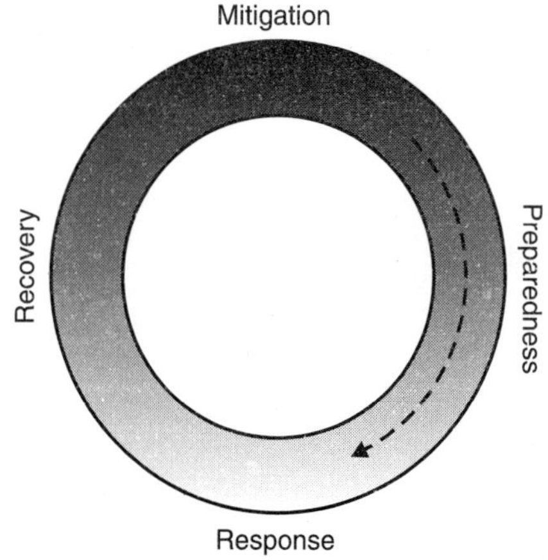

Fig. 9.1. A graphic rerpresentation of four phases of Disaster management.

(1) Mitigation (2) Preparedness

(3) Response (4) Recovery

(1) Mitigation : Mitigation efforts prevent hazards from developing into disasters or to reduce the effects of disasters when they occur. Example of mitigation are strengthening and retrofitting of buildings to make them earthquake resistance, creating awareness in public about earthquakes through media and schools so that they should know what to do during earthquakes. It also include enforcement of building codes and their revision from time to time.

(2) Preparedness : In the preparedness phase, emergency managers develop plans of action for when the disaster strikes. It include the proper training programmes for rescue workers, maintenance and checking of equipments and emergency services, evacuation plans and control of supplies etc. It also include developing volunteers from common people to work during rescue operations.

(3) Response : The response phase is the reflection of country's readiness to cope quickly and effectively during and after disaster. It include the mobilization of necessary emergency services like police, volunteers, fire fighters and NGO's. A well practised plan developed in preparedness phase can result in an effective rescue efforts. Rehersals of rescue plans is essential to achieve optimum output in the response phase, medical facilities should be checked and enhanced from time to time. The response phase start with rescue phase which is discussed later. For example : to facilitate emergency operations, the earthquake prone regions should be equipped with a number of emergency centres. The role of such centres should be to maintain stock of food, medicine, building material etc. for emergencies. They should also do the task of survey, assessment of damages and teaching of evacuation measures to common people.

(4) Recovery : The aim of the recovery phase is to restore the affected area. Recovery phase is mainly concerned with rebuilding destroyed property, re-employment, repair of other essential infrastructures and services. The recovery phase starts when the immediate threat to human life has reduced. In the long term disasters, like wars, famine etc. which may last up to many months, the recovery phase starts early. The recovery phase is generally long and may take upto five years. There are three main categories of recovery phase.

(*a*) **Restoration :** Restoration of essential services, homes and other infrastructure.

(*b*) **Rehabilitation :** Physical and mental treatment of affected people.

(*c*) **Reconstruction :** Reconstruction of collapsed buildings and infrastructure.

All these phases should follow one after the other and the cycle should be completed for effective and result oriented disaster management programme.

9.3. DISASTER RESCUE

The aim of rescue is to save life and minimize injury to people and damage to property. Rescue involves many functions. Some of the common rescue functions are

(*a*) To remove trapped people and support them during rescue operations.

(*b*) To recover and identify the dead.

(*c*) To provide support to other services or teams.

In order to achieve the aim of rescue all rescuers should be trained properly.

9.4. PSYCHOLOGY OF RESCUE

Rescue operation is also full of danger. People tend to react differently to danger. But mostly there is a feeling of anxiety and fear in all the persons because it is not just the victim who faces the danger, but the rescuer also face the same danger in order to rescue the victim. Dangers are still present when the main danger has passed. The main difference between the condition of rescuer and victim is that the rescuer is better able to handle the situation because he can for see the danger and he is well equipped. It is normal to be anxious and feel fear during danger. Other emotions which arise during a rescue operation are pity, disgust, contempt, pride, concern and many more. The pressure of urgency or emergency is always their on rescuer. It decreases the efficiency of the rescue operations. The rescuer must understand the psychology of victims also in addition to the physical needs. He should help them in coming out of the trauma.

9.5. RESCUE WORKERS

Rescue workers are those persons who are involved in rescue operations. Mainly three types of rescue workers are there

(1) **Survivors :** Survivors are those persons who have escaped in the danger i.e. they are not injured. The first reaction of survivors, is to help their neighbours and families. Sometimes, they don't know what to do but feel that they must do something. Their intention is good but sometimes it hamper or interrupt the function of trained rescue teams. But, survivors are the first one to help victims. For example, in case of fire or toxic gas exist at the site of emergency, the first group to start rescue work is of the survivors. But there is danger also involve with this group of rescue workers because they are untrained.

(2) **Untrained People :** The second group of rescue workers are those people who witness the event or are in the neighbourhood of the event. Some of these people have a desire to help the victims. They also bring necessary things with them and can be very effective if these things are properly supervised. But a large number of people are curious and just want to watch without helping rescuers and victims. They can cause interference in rescue operations and should be controlled.

(3) **Trained People :** These are the last one to arrive at the scene. These are the trained rescuers from police, fire NGO's and state emergency departments. These people know what to do and how to use the available resources and materials. They can also take the help of untrained people efficiently to carry out rescue operations in proper way.

9.5.1. Qualities of the Rescuer

The Rescue operation is a very difficult task. All people are not suited to such work. There are some minimum qualities which a rescuer must possess. These qualities are listed below :

(*a*) **Physical fitness :** The rescue operation require stamina and physical fitness because they often continue for long time and in challenging environment.

(*b*) **Interest :** The rescuer should have genuine interest in rescue work. He should not have any feeling of impressing people etc.

(*c*) **Training :** The rescuer should have undergone training which will help him in performing the rescue work efficiently and in safe manner.

(*d*) **Initiative :** The rescue operation is a very big and complex process. Thus, it is impossible to closely supervise each team member. Each must be able to see and plan what is needed at the movement and do accordingly.

(*e*) **Versatility :** Each situation or emergency is one of its kind. So, the rescuer must be able to understand the problems and use the skills and knowledge to handle new situations.

(*f*) **Co-operation :** Rescue work involves lot of people. To make it successful, a good team effort is needed. Thus, co-operation with other team members is an important quality of the rescuer.

(*g*) **Leadership qualities :** All rescuers should have leadership qualities. They require it at various times. A rescuer having leadership qualities can make proper and safe use of many more untrained people.

(*h*) **Controlled fears :** Rescuers must know what are his limitations. He should be aware of any fears he is having and should be able to control them. It is also important that the leader of a rescue team should know if any team member has fear of any kind. Some fears which commonly affect a rescuer should be identified and explained in the training programmes.

(*i*) **Confidence :** A rescuer must have confidence in himself and his acts. His appearance must instill faith and confidence in others also.

9.6. RESCUE PLAN

The rescue operation should be performed in a planned way. The success of rescue operations depends on the planning of it. The rescue plan consists of following major stages.

(*a*) The Reconnaissance : Every member of the rescue team must be trained in rescue reconnaissance. In most of the cases team leader does the reconnaissance but if he has deployed some person for conducting reconnaissance, then the person should take and report the observations back to the leader. The reconnaissance of the disaster area/site should result in an accurate assessment of :

1. The number and location of casualties.
2. Dangerous situations such as gas, leak electricity, overhanging walls, unsafe structural components or anything else which may endanger rescue workers or survivors.
3. Access to the casualties.
4. The extent and type of damage.
5. Emergency services available and disrupted.
6. Available resources, both personnel and equipment.
7. An approximate analysis of the time, the task will take with available resources.
8. Support agencies.

(*b*) Problem Solving Process : This is also called as appreciation process. It is a simple method of problem solving which involves logical assessment of the situation and the reconnaissance. It results in the formation of a workable plan. Experienced persons may use their experience and intuition also to make the rescue plan more effective. The problem solving process has six steps :

1. **Define the problem :** The problem to be solved should be clearly defined in terms of the task which is to be performed. If the problem is too large or complex to tackle easily, then it may be divided into a number of small elements, which are connected properly.
2. **State the aim :** The aim is a clear picture of what the rescue team has to do in order to solve the problem. The aim should be clear brief and achievable. For example, one of the aim during an earthquake rescue operation can be, "to rescue the casualty from the basement floor".
3. **Consider the factors :** The factors to be considered in a rescue operation may include :
   (*i*) Number and location of casualties
   (*ii*) Time
   (*iii*) Topography of the area
   (*iv*) Weather
   (*v*) Available resources, both personnel and equipment
   (*vi*) Support requirements and availability
   (*vii*) Communication
   (*viii*) Priority of tasks

   Each factor should be considered and all the "ifs" and "thens" should be analysed properly. Each factor should lead to some conclusions. For example :

   Factor : A persons legs are trapped under a heavy steel beam.

   Conclusion : The rescue team must use cutting and lifting equipment carefully to free the person.
4. **Determine all possible solutions :** Consider all possible solutions, that are practical, for achieving the aim. This is done by considering all the factors listed above.
5. **Decide on the best possible solution :** At this stage one of the best possible solution is selected which takes into account all the factors. It is the solution with the most advantages and least disadvantages.
6. **Plan :** On the basis of best possible solution the plan is developed. The plan must be simple and it must relate directly to the aim. When the rescue plan is completed it should be checked thoroughly with proper reasoning.

After the rescue plan is prepared the rescue team should start the action without delay. The team leader should decide and deploy persons for various tasks and give priorities to them. For example, Rescue workers deployed on a collapsed building should carefully observe how the building has collapsed. He should identify and explore the debris, voids etc. which can be used to access the casualties. During disaster management, a leader has to use his experience and training and combine them with effective decision making.

9.7. RESCUE BY STEPS

A six phase process for rescue operation at a collapsed structure is given below which explains the rescue operation by steps. These six phases are recognised internationally and are used by rescuers worldwide.

Phase 1 : **R**econnaissance and Survey

Phase 2 : **E**limination of utilities

Phase 3 : **P**rimary surface search and rescue

Phase 4 : **E**xploration of all voids

Phase 5 : **A**ccess by selected debris removal

Phase 6 : **T**erminate by general debris removal.

The REPEAT acronym helps the rescuers in learning and remembering these six phases :

Phase 1 : Reconnaissance and Survey : After reaching the hazard site, it is difficult to conduct thorough reconnaissance and survey. During reconnaissance, question victims, witnesses and others who are able to provide information such as occupancy of the building, location of victims and potential hazards etc. Survey involves a visual inspection of the site and may identify the type and size of the structure, collapse patterns, possible hazards and location of victims etc.

Phase 2 : Elimination of utilities (Risk Assessment and Control) : It is one of the most important phases. It involves the assessment of risks and controlling them with the measures available to the rescuers. For example barricading the site or shutting of power supply etc. It is necessary to control all the possible risks as soon as possible otherwise it may result in more disaster.

Phase 3 : Primary Surface Search and Rescue : The third phase include searching surface areas around the collapsed or damaged structure. Any injured person or lightly trapped person should be immediately attended. During the primary surface search, rescuers may become aware of victims located within voids or within damaged structures. The location of victims should be marked and reported to trained persons with appropriate equipment for safely extraction of victims.

Phase 4 : Exploration of all Voids : In this phase, specially trained and equipped rescuers explore those places which have been marked during primary surface search. All the places, where the trapped victims might have survived the collapses, are explored although there is no evidence of the presence of such victims. If possible, technical devices can also be used to know the location of the trapped victims.

Phase 5 : Access by Selected Debris Removal : Phase 5 involves the removal of selected debris for accessing the trapped victims. This require specially trained personnels and equipments. As the time passes, the chances of survival of victims decreases. Thus, the rescue operation should be carried out speedily but safely. It is essential to locate all victims before starting phase 6.

Phase 6 : Terminate by General Debris Removal : Phase 6 involves the use of heavy plant and machinery to remove all the debris in an attempt to recover and account for all victims. This phase also involves the use of forensic processes to identify victims. It is important to note that the rescue operation must be completed without interruption until phase 6 is completed. Every person known or likely to have been in the structure must be accounted for. It is impossible to tell from external inspection whether victims buried in the debris will be alive or not. Thus the effort should be continued to save lives, no matter how long it take.

9.8. RESCUE EQUIPMENT

The rescue operation require lot of equipments. The equipment to be used depend upon the type of task involved. It is necessary to check the equipment before and after use. It should be in a proper working condition. Because the non-working of any equipment may cause unnecessary delay in rescue work. For example, the fuel driven vehicle or equipment should have enough fuel for non-stop working. These equipments should be operated by trained persons as per the instructions. Some of the equipments used for rescue operation after an earthquake are :

(1) JCB/bull dozers
(2) Excavators
(3) Dumpers
(4) Cranes
(5) Air compressors
(6) Gas cutter
(7) Generators
(8) Pumps
(9) Fire extinguishers
(10) Ropes and chains
(11) Ladders
(12) Hose pipe

9.9. SAFETY IN RESCUE OPERATIONS

The rescue operation involve lot of works which need skills like clearing of debris, operation of equipments, working in dangerous environment like fire and collapsed structure etc. These tasks if not properly carried out, may prove dangerous. Thus safety in rescue operations is of main importance. It is necessary to pay attention to safety measures used in rescue operations. It is also needed to observe and follow these safety measures strictly during the rescue operation, training and exercises also. Most of these safety precaution are very basic and simple.

9.9.1. Basic Precautions

- Regularly and carefully check the equipment both before and after the use.
- Ropes can wear and rot, batteries can corrode equipment, machinery can break down.
- Faulty equipment can cost lives. Immediately label the faulty equipment and send it for repair or replace it.
- Protect people at heights or depths with care and monitoring.
- Never ignore the safety limits.
- Control the emotions during rescue operations.

These basic precautions help in protecting the casualty and the rescuer. Other types of safety measures are explained in the following articles.

9.9.2. Rescue Worker Safety

Each rescuer should be provided with protective clothing, safety boots and helmets which must be properly used

- Wear helmets at all the times of risk.
- Gloves should be used while clearing debris with hands.
- Dust mask or filter should be of standard quality and to be used when required.
- Safety goggles should be used for protecting eyes.
- Clothing should be protective according to the hazard (fire, flood etc.).
- Safety boots and knee or elbow protection when moving over debris.
- Sun screen and hat for working in the hot sun.
- Water to prevent dehydration.
- Whistle is to be provided for emergency times.

9.9.3. Casualty Safety

The safety of casualties is of prime importance. Every effort is to be made to ensure that no harm should be done to the casualty during rescue operation.

- The injured persons should be given care and first aid immediately.
- They should be handled in a safer way and made comfortable.
- The damaged building parts or the debris should be handled such that no more injury is done to the people trapped inside.
- Priority attention is given to the casualties depending on their condition.
- The safety measures should be discussed during the training and exercises along with the dummy casualties.
- The injured persons are sent to the hospitals by the ambulances or other vehicles. These vehicles should be driven by trained drivers.

9.9.4. Equipment Safety

- All equipments should be of standard quality and operated as per instructions. They should be operated by trained persons.
- While using power tools such as disc cutters or saws, the operator should wear safety goggles and gloves and take proper care to avoid jamming of blades.
- If the cutting operation is near the trapped casualty, the cutting blades should be directed away from the casualty body.
- Proper care is to be taken while load lifting operations to prevent total collapse or slippage or dropping.
- All rescue equipment should be checked before and after use.

9.9.5. Hazards Arise Due to Breakdown of Public Utilities

The disaster can lead to breakdown of public services like leakage of Gas or water services or bring down electrical power lives or may cause fire. All utility services should be treated with care. The casualty should be removed immediately from the dangerous area. In a disaster, the rescuer must consider their own safety also important.

(*a*) **Leakage of gas :** Leakage of gas may cause explosion thus safety precautions must be followed during this hazard.

(*i*) Turn off the supply and allow time for the gas to clear.

(*ii*) Don't use match stick. Use soapy water to find a leak.

(*iii*) Don't switch on or off any electrical appliances.

(*iv*) If a fire is present use water spray and call fire extinguishers.

(*b*) **Breaking of Water Pipes :** Water from broken pipes may enter areas in which casualties are trapped. Steps should be taken to reduce the risk of immersion or drowning.

(*c*) **Broken Sewers :** Broken sewers may create problems of flooding and escaping gases sewer gases can be explosive as well as toxic. Thus, following precautions are to be followed during this hazard.

- Take proper action (use of mask etc.) before entering the area.
- Never use an open flame.
- Pumping can be used to divert the flow away.

(*d*) **Breaking of electric wires :**

- Do not touch the electric wires
- Avoid water near the electric wires, avoid other conductors also.
- Switch off the electric supply to a damaged building.
- Be careful at night, when it is difficult to see wires.

9.10. DEBRIS CLEARANCE

In a disaster, where collapse or damage of structure is there, the people are usually buried or trapped inside the debris. Thus, for extracting the people out of debris, the debris are to be removed or cleared. The debris are to be disturbed as little as possible during rescue operations to avoid internal collapse. It is also to be made sure that after removing a part of debris, the remainder should not slide or fall. Clearance of debris require special methods to avoid injury to trapped people.

There is a difference between the debris clearance as a rescue operation and debris clearance to clear a site. The first case is very slow and utmost care is required as our aim is to extract the victims. While, the later case can be very speedy depending upon the manpower and equipment.

9.10.1. Method of Debris Clearance

When debris clearance is required for rescue purposes, move the debris clear of the collapsed building. Remove debris by hand, piece by piece until the casualties are uncovered and freed. It is best to form a human chain. Sometimes it is necessary to cut a lane through the debris to reach a casualty.

Great care should be taken in doing so and it should be ensured that the sides of the lane do not collapse. Timbering and strutting can be done, where necessary to prevent the collapse of the sides.

9.10.2. Precautions in Debris Clearance

Debric removal is a very important step in rescue operation. If performed properly, lot of persons can be saved. The following precautions should be observed while clearing debris, for rescuing people

(*a*) Identify suitable escape paths and keep them clear.

(*b*) Use the edged tools very carefully, while removing debris.

(*c*) Remove debris close to casualties by hand.

(*d*) Wear gloves

(*e*) Do not climb over debris during the clearing operation unless absolutely necessary.

(*f*) Remove debris only, when it is certain that no further collapse will be caused.

(*g*) Operate heavy equipment only as per instructions.

(*h*) Carefully observe the movement of damaged parts of building, while removing debris.

9.11. CASUALTY MANAGEMENT

In a disaster like earthquake, the rescue operation is difficult and long. The people are generally trapped inside the debris. They are to be taken out very carefully. The method used for casualty removal depend on the location of the casualty and the type of injury sustained. In some rescue operations, casualties will have to be lowered from the upper floors of buildings. Casualities are to be handled carefully by rescue workers so that more injuries does not occur. Their safety is very important. All the rescue workers should be given First Aid training and they should be well equipped to give first aid to the injured persons.

After removal, many casualties will have to be carried over pile of debris and uneven ground before handed over to the ambulance. This operation should be smooth and done carefully to avoid any injury or shock to the casualty. Speed of removal is important but safety and proper handling is also important. Rescue leader should conduct exercises in the removal of casualties, using live people as casualties. This will help the rescue workers to understand the problems faced while handling the casualties.

9.11.1. Triage

When the number of casualties is large, the casualties are rescued and treated by order of priority. This selection or sorting of priorities is called as triage. The rescuer has to perform the task of triage. While performing triage the rescuer must perform the following tasks carefully.

(*a*) Quickly assess the condition of each casualty.

(*b*) Place unconscious person in a position so that they can breathe properly.

(*c*) Try to control bleeding temporarily

(*d*) Select the casualties with several injuries on a priority basis.

9.11.2. Classification of Casualties

Casualities are classified into following four groups :

(*a*) **Severely Injured :** There are the injured persons who need urgent medical care. They should be given first priority. Severely injured persons require air way, breathing and transportation support. Trained persons should assist them and proper medical care is given to them immediately.

(*b*) **Significantly Injured :** These are the injured person whose condition is stable and treatment can wait. Casualties with significant injuries need medical care and hospitalisation. However, their condition is such that they will survive even if the treatment is delayed.

(*c*) **Minor Injuries :** These are the persons who are wounded but are able to move. These are not to be hospitalised. Some of the persons who are psychologically disturbed due to trauma also come in this category. This group is referred as "walking wounded", but they should not be allowed to walk if they show signs of shock or internal injuries.

(*d*) **Deceased :** These are the dead persons. The medical officers should certify their death. These casualties should not be moved unless it is necessary to access live casualties.

TERMS TO REMEMBER

1. **Disaster :** A disaster is a phenomenon which puts humans and property in a dangerous or disadvantageous situation.
2. **Disaster management :** It is the discipline dealing with and avoiding risk caused due to disasters.
3. **Rescue :** It is the process/operation to save life and minimizing injury to people and damage to property.
4. **Rescue workers :** Rescue workers are those people who are involved in rescue operation. They include common people and trained people.
5. **Rescue Plan :** A series of well co-ordinated and workable activities for saving the lives of people after the disaster.
6. **Casualty :** The people who are affected from the disaster.

EXERCISES

A. Objective Type Questions :

1. A disaster can be
   (*a*) Natural (*b*) Man-made
   (*c*) (*a*) and (*b*) (*d*) None of the above
2. Disaster management has following four phases in a sequence
   (*a*) Mitigation, preparedness, response, recovery
   (*b*) Response, mitigation, recovery, preparedness
   (*c*) Mitigation, response, preparedness, recovery
   (*d*) Recovery, response, preparedness, recovery.
3. ________ refers to the preventing hazards from developing into disasters.
   (*a*) Response (*b*) Mitigation
   (*c*) Preparedness (*d*) Recovery
4. The rescue phase comes under
   (*a*) Mitigation (*b*) Preparedness
   (*c*) Response (*d*) Recovery
5. The following phase refers to the restoration and rehabilitation
   (*a*) Mitigation (*b*) Preparedness
   (*c*) Response (*d*) Recovery
6. Debris clearance, near the trapped persons should be done
   (*a*) Carefully removing pieces one by one by hand.
   (*b*) By removing debris with equipment.
   (*c*) By blasting the debris
   (*d*) None of the above.
7. The selection or sorting or giving priorities to casualties is called as ________.
   (*a*) Trial (*b*) Triage
   (*c*) Tryed (*d*) Triad
8. Severely injured persons should be attended
   (*a*) Last (*b*) Second
   (*c*) First (*d*) Need not be attended
9. Rescue by steps has got _______ phases.
   (*a*) Eight (*b*) Six
   (*c*) Four (*d*) Five
10. Reconnaissance is the _______ step in rescue planning.
    (*a*) Last (*b*) Second
    (*c*) First (*d*) None of the above.

B. Fill in the Blanks :

1. The disaster can be _________ or _________.
2. _________ helps in reducing the damage and risks caused by disasters.
3. The most important activity after disaster occurrence is _________.
4. Strengthening and retrofitting is the _________ aspect of disaster management.

5. Mitigation helps in _________ the effects of disaster.
6. _________ phase is to restore the affected area.
7. The aim of _________ is to save life and minimize the damage.
8. The four phases of disaster management are _________, _________, _________ and _________.
9. The longest phase of disaster management is _________.
10. _________ are those people who have escaped in danger.
11. NGO's are the _________ rescuer.
12. The first stage of rescue plan is _________.
13. Problem solving process is also called as _________ process.
14. The selection or sorting of casualties to give priority is called as _________.
15. Deceased are the _________ persons.
16. Severely injured persons need _________ medical care.
17. People who are affected from disaster are called as _________.
18. A proper and well co-ordinated rescue plan can _________ many lives.
19. The rescue equipment should be used by _________ person.
20. _________ are those people who are involved in rescue operation.

C. State True/False :

1. All disasters are naturally occurring.
2. Cyclone is a wind related disaster.
3. Mitigation prevents hazards from turning into disaster.
4. Strengthening and retrofitting a structure is a part of recovery phase.
5. Rescue operation is full of danger.
6. Assessment of risks is not a part of rescue operation.
7. All rescue workers must be given first aid training.
8. Significantly injured are those persons whose condition is not stable.
9. Walking wounded are the minorly injured persons.
10. In the collapse of a building the debris must be removed quickly.

ANSWERS

A. Objective Type Questions :

1. (*c*)	**2.** (*a*)	**3.** (*b*)	**4.** (*c*)	**5.** (*d*)
6. (*a*)	**7.** (*b*)	**8.** (*c*)	**9.** (*b*)	**10.** (*c*)

B. Fill in the Blanks :

1. natural, man made **2.** Disaster management
3. Rescue operation **4.** Mitigation **5.** reducing
6. Recovery **7.** Rescue
8. Mitigation, preparedness, response, recovery
9. Recovery **10.** Survivors **11.** trained
12. Reconnaissance **13.** Appreciation **14.** Triage
15. dead **16.** urgent **17.** Casualty **18.** save
19. trained **20.** Rescue workers.

C. State True/False :

1. False	2. True	3. True	4. False	5. True
6. False	7. True	8. False	9. True	10. False

SHORT ANSWER TYPE QUESTIONS

1. What do you understand by the term disaster ?
2. What are the four phases of disaster management.
3. Who is a rescue worker ? Name the types of rescue workers.
4. Define rescue plan.
5. Enumerate some natural disasters.
6. What is the aim of rescue operation ?
7. What are the major stages of a rescue plan ?
8. List six phases of the rescue operation at a collapsed building.
9. Name some of the equipments used in rescue operation.
10. What is casualty ? Name its types in the order of giving priority during rescue operation.
11. Define triage. What is the basis of performing triage ?
12. What do you understand by the term 'casualty management' ?

LONG ANSWER TYPE QUESTIONS

1. What is disaster management ? Explain its various phases.
2. What do you understand by term "Disaster Rescue". Explain the psychology of rescue.
3. What are the types of rescue workers. Explain their role in rescue operation.
4. Explain the qualities of a rescue worker.
5. What is "Rescue Plan" ? What are the processes required for obtaining an efficient rescue plan.
6. What is rescue by steps ? List various phases of it.
7. Explain various phases of 'Rescue by Steps'.
8. What are the various rescue equipments used in rescue operations.
9. What is the importance of safety operations in rescue operations.
10. What are the basic precautions to be followed in rescue operations.
11. Explain various types of safety measures to be followed in rescue operations.
12. What are the safety precautions while using equipments and vehicles by rescue workers.
13. What are the safety measures to be followed for preventing hazards due to breakdown of public utilities.
14. Explain the method of debris clearance. What are the precautions to be followed in debris clearance ?
15. What are the different types of casualties. What are the points to be kept in mind while managing casualties.

❑❑❑

Annexure

Special Construction Methodologies-Smart Buildings

Common seismic design attempts to make buildings that do not collapse under strong earthquake but may get demaged. This may make the building non-functional. In some important buildings like hospital, it may prove disaster, thus, they need to remain functional all the time. Special techniques are used to design these important buildings now-a-days so that they remain practically undamged even in a severe earthquake. The cost of construction of these buildings is higher than normal buildings but the cost is justified for its usefulness. Two basic methods are used to protect the buildings from damaging earthquake effects.

1. Base Isolation Devices
2. Seismic Dampers

1. Base Isolation : The concept of protecting a building from the damaging effects of an earthquake by introducing some type of support that isolates it from the shaking of ground. In this some isolation system mainly bearings are introduced between the building and the foundation. When the ground shakes, the isolation bearings will move freely but the building will not move. Thus no force is transferred to the building due to shaking of the ground or simply, the building does not experience the earthquake. The following figure explains the concept of base isolation in a systematic way.

The most common base isolators are short cylinderical bearings with alternating layers of steel plates and hard rubber pads between the base of the building and the foundation. This system makes the building flexible and absorb energy also. Thus help in reducing the seismic effects.

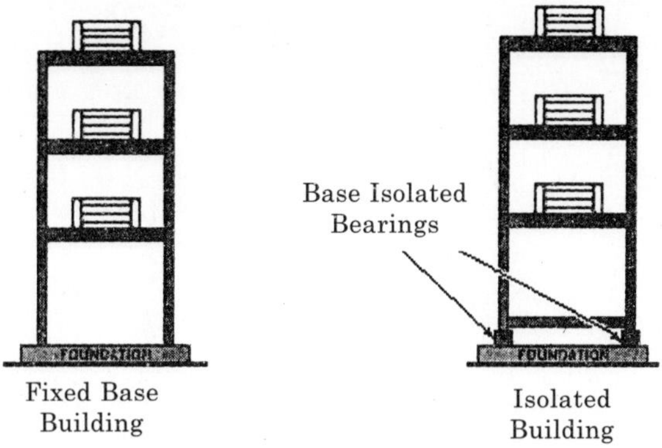

(A)

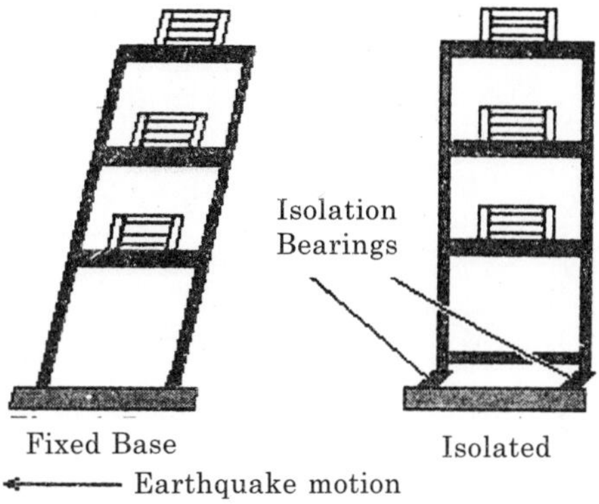

(B)

2. Seismic Dampers : Another method for controlling seismic damage in buildings and improving their seismic performance is by installing seismic dampers in place of structural braces. These dampers act like shock absorbers of cars-which absorbs shocks or sudden jerks. These damp the motion of the building thus reducing the seismic effects. Commonly used seismic dampers are viscous dampers, friction dampers and yielding dampers. In India, friction dampers have been provided in a 18 storeyed RC frame structure in Gurgaon, Haryana.

❑❑❑

MODEL QUESTION PAPERS

MODEL QUESTION PAPER 1

EARTHQUAKE RESISTANT BUILDING CONSTRUCTION

Time : 3 Hours *Total Marks : 75*

Note : Attempt all questions.

Q. 1. Fill in the blanks : **(10)**

(*i*) The upper most layer of the earth is called as ________.

(*ii*) Shimla comes under zone ________.

(*iii*) The earthquakes which occur along the tectonic plate boundaries are ________.

(*iv*) The most common mode of failure of masonry buildings is ________.

(*v*) S waves can travel through ________.

(*vi*) Liquefaction is a phenomenon commonly seen in ________ soils.

(*vii*) Non engineered construction are called as ________ construction.

(*viii*) Open ground storey is also called as ________ storey.

(*ix*) Minimum grade of concrete for R.C.C., as per IS : 456-2000 is ________.

(*x*) The seismic code IS ________ deals with general provisions for buildings.

Q. 2. State True or False : **(5)**

(*i*) Record of an earthquake is called as seismograph.

(*ii*) Study of earthquakes is called as seismology.

(*iii*) Symmetrical buildings are subjected to more damage during earthquakes.

(*iv*) Loads which changes with time are dynamic loads.

(*v*) Bending failure is also known as out of plane failure.

Q. 3. Explain any ten out of following : **(10 × 3)**

(*i*) Isoseismals

(*ii*) Seismograph

(*iii*) Epicenter

(*iv*) Diaphragm failure

(*v*) Floating columns

(*vi*) Seismic zoning

(*vii*) 'through' stone used in stone masonry

(*viii*) Pounding

(*ix*) Shear walls

(*x*) Box action

(*xi*) Inertia force

(*xii*) Soft storey

Q. 4. Attempt any three questions. **(3 × 10)**

(*i*) What are the causes of earthquakes ? What is the difference between magnitude and intensity.

(*ii*) Explain the following modes of failure observed in a masonry building.

(*a*) Out of plane failure

(*b*) In plane failure.

(*iii*) What are seismic waves ? Explain any two types of waves.

(*iv*) Explain the commonly observed damages seen in columns and beams of RCC buildings.

(*v*) Explain the guidelines for earthquake resistant design and construction of stone masonry buildings.

MODEL QUESTION PAPER 2

EARTHQUAKE RESISTANT BUILDING CONSTRUCTION

Time : 3 Hours *Total Marks : 75*

Note : Attempt all questions.

Q. 1. Fill in the blanks : **(10)**

(*i*) Earthquake intensity is measured on ________ scale.

(*ii*) ________ is the body wave in which material particles vibrate at right angle to the direction of propagation.

(*iii*) ________ is the record of seismic waves.

(*iv*) The fundamental natural period of flexible buildings is ________ than 1.0 seconds.

(*v*) ________ earthquake occurred in 2001 in Gujarat.

(*vi*) In plane failure is commonly seen in ________ walls.

(*vii*) Delhi lies in zone ________ as per IS : 1893 : 2002 (part I).

(*viii*) ________ is the property in which material undergoes appreciable deformation before failure.

(*ix*) Floating column is an example of ________ irregularity.

(*x*) ________ is the code used for ductile detailing of structures.

Q. 2. State True or False : **(5)**

(*i*) In fill walls are load bearing walls.

(*ii*) A regular and symmetric building has very good earthquake resistance.

(*iii*) Soft storey buildings are prone to more damage during earthquakes.

(*iv*) The center to center spacing of lateral ties in columns near the ends should be more.

(*v*) Seismic shear force is more for heavier buildings.

Q. 3. Explain the following terms (any ten) **(10 × 3)**

(*i*) Faults

(*ii*) Focal depth

(*iii*) Seismoscope

(*iv*) Dynamic loads

(*v*) Pounding

(*vi*) Liquefaction

(*vii*) Seismic zone factor

(*viii*) Short column effect

(*ix*) Framed construction

(*x*) Connection failure

(*xi*) Mass irregularity

(*xii*) Natural period.

Q. 4. Answer any three questions in detail. **(3 × 10)**

(*i*) Discuss the classification of earthquake.

(*ii*) Explain any two modes of failure of masonry buildings during earthquake.

(*iii*) Explain the seismic strengthening arrangements (bands) of masonry buildings.

(*iv*) Explain any five general principles used in the construction of earthquake resistant structures.

(*v*) What do you understand by the term "Irregularity" in buildings. What are its types. Explain any one type of irregularity.

MODEL QUESTION PAPER 3

EARTHQUAKE RESISTANT BUILDING CONSTRUCTION

Time : 3 Hours *Total Marks : 75*

Note : Attempt all questions.

Q. 1. Fill in the blanks : **(10)**

(*i*) ________ is the branch of science that deals with the study of earthquakes.

(*ii*) Intensity of vibratins ________ with increase in epicentral distance.

(*iii*) ________ refers to giant sea waves.

(*iv*) The tendency of a body to resist the change of state is called as ________.

(*v*) Reinforced concrete or reinforced brick slabs act as ________ diaphragms.

(*vi*) Openings in a wall should be placed ________ from the corners.

(*vii*) IS 1893 : 2002 (part 1) deals with ________.

(*viii*) For good earthquake resistance, the columns of a building should be made ________ than beams.

(*ix*) ________ prevent buckling of the vertical bars of column.

(*x*) The masonry walls used to fill spaces are called as ________.

Q. 2. State True or False : **(5)**

(*i*) The record of an earthquake is called as seismograph.

(*ii*) P waves are faster than S waves.

(*iii*) Bending failure in masonry is observed by diagonal cracks.

(*iv*) The roof covering should be made of light material.

(*v*) Open ground storey is called as strong storey.

Q. 3. Explain the following terms (any five) **(5 × 3)**

(*i*) Epicenter

(*ii*) Isoseismal

(*iii*) Mass irregularity

(*iv*) Floating column

(*v*) Shallow earthquake

(*vi*) Box action

(*vii*) Wythes

Q. 4. Give reasons for following (any five) **(5 × 3)**

(*i*) Collapse of masonry structures has resulted in heavy damage and loss of lives.

(*ii*) Foundation should not be constructed on loose soils.

(*iii*) The walls of a building should be properly tied or bonded together.

(*iv*) Adjoining buildings should have proper separation between them.

(*v*) Openings in walls should be small and placed away from the corners.

(*vi*) It is advisable to have separate small rooms than one long room in a building.

(*vii*) Bands are provided in masonry buildings.

Q. 5. Answer in detail any three **(3 × 10)**

(*i*) Explain the causes of earthquakes. Why is it important to study earthquakes engg. for a civil engineer ?

(*ii*) Explain various causes of damage and failure of masonry buildings.

(*iii*) Discuss the following modes of failure and their prevention.

(*a*) Diaphragm failure

(*b*) Connection failure

(*iv*) Explain with the help of a diagram, various codal provisions for ductile detailing of reinforcement in a R.C.C. beam.

(*v*) Discuss any four types of commonly observed damages in R.C.C. buildings.

MODEL QUESTION PAPER 4

EARTHQUAKE RESISTANT BUILDING CONSTRUCTION

Time : 3 Hours *Total Marks : 100*

Note : (*i*) Question No. 1 is compulsory attempt any four from the rest questions.

(*ii*) Use of IS : 1893 (Part) : 2002 is permitted.

1. Explain clearly with reasons (any four) **(5 × 4 = 20)**

(*a*) Well connected masonry walls show better seismic performance.

(*b*) Out of plane failure is more common than in plane failure.

(*c*) A regular symmetric and simple building shows good seismic performance.

(*d*) Provision of "through stone" in stone masonry walls for seismic strengthening.

(*e*) Open ground storeys are liable to fail during an earthquake.

(*f*) Disaster management can save many lives.

2. (*a*) What are the ways in which earthquakes can be classified ? **(20)**

(*b*) What are the different types of seismic waves ? Which waves causes maximum destruction during earthquakes ?

3. (*a*) What are the relative advantages and disadvantages of masonry constructions over other types of constructions ? **(20)**

(*b*) What are the various seismic strengthening arrangements of a masonry building ? Explain them clearly with the help of a diagram.

4. (*a*) What are the common seismic damages seen in masonry buildings. Explain them alongwith their causes.

(*b*) What is a short column effect. What are the special ductile detailing requirements for a short column. **(20)**

5. A four storeyed RC building (ordinary moment resisting frame) has a frame and plan shown in Fig. 1. Determine the lateral forces caused due to earthquake as per IS code method. **(20)**

Date Given :	Place	Delhi
	Floor height	3.5 m
	Brick walls	250 mm thick including plaster
	Live load	4.0 KN/m^2
	M20 Concrete and Fe 415 steel	
	Columns	250 × 500 mm

Beams	250 × 400 mm
Slab	100 mm thick
Soil	Rock

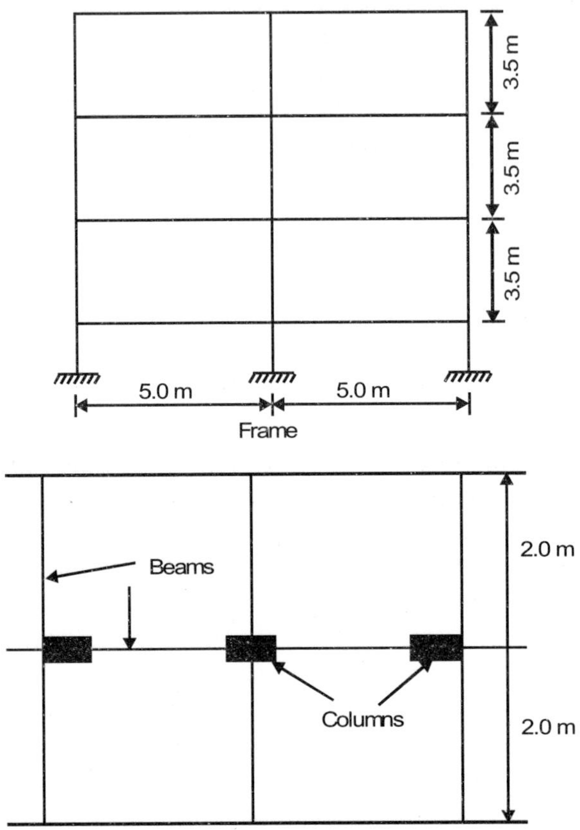

Fig. 1.

6. (*a*) What do you understand by the term disaster management ? What are its various phases ? **(20)**

   (*b*) Explain the various guidelines to be followed while planning, designing and construction of earthquake resistant buildings.

MODEL QUESTION PAPER 5

EARTHQUAKE RESISTANT BUILDING CONSTRUCTION

Time : 3 Hours *Total Marks : 100*

Note : Attempt any five questions.

1. (*a*) What is seismology ? Explain various types of seismic waves.
   (*b*) What is meant by earthquake size ? Distinguish clearly between magnitude and intensity of an earthquake. **(20)**
2. (*a*) What is an isoseismal map ? What is the use of seismic zoning ?
   (*b*) State the equation of dynamic equilibrium. Exaplain it in relation with earthquake motion. **(20)**
3. (*a*) What are the tradionally built constructions ?
   (*b*) Explain clearly with the help of diagrams various modes of failure of a masonry building. **(20)**
4. (*a*) Explain the various principles to be observed while designing earthquake resistance buildings.
   (*b*) What is a horizontal band ? What are the various types of bands provided in masonry buildings ? **(20)**
5. (*a*) Explain various earthquake resistant features to be provided in stone masonry buildings.
   (*b*) Discuss in detail the various guidelines for earthquake resistant construction of earthen buildings. **(20)**
6. (*a*) What do you mean by irregular building ? What is the effect of irregularity on seismic performance of a building. **(20)**
   (*b*) Explain various types of vertical irregularities in a building. **(20)**
7. (*a*) What are the various seismic codes published by Bureau of Indian Standard.
   (*b*) Discuss various types of seismic damages seen in columns and joints of a reinforced concrete building. **(20)**
8. Write short notes on any four of the following : **(20)**
   (*a*) Rescue by steps.
   (*b*) Classification of earthquakes.
   (*c*) Ductile detailing of lateral ties in a column.
   (*d*) Casualty management.
   (*e*) Fundamental natural period.
   (*f*) Seismic design philosphy.

❑❑❑

EXAMINATION PAPER

EXAMINATION PAPER

EARTHQUAKE RESISTANT BUILDING CONSTRUCTION

SIXTH EXAM/CIVIL/5132/6851/May., 2016

Duration : 3 Hrs. *M.Marks: 75*

SECTION-A

1. Fill in the blanks : **(10×1=10)**

(*i*) The point where earthquake originates is known as ________.

(*ii*) Focus is also known as ________.

(*iii*) The record of earthquake is known as ________.

(*iv*) Primary and secondary waves are also known as ________.

(*v*) Most of earthquake originates due to ________ cause.

(*vi*) The fundamental period of 15 storey building is ________.

(*vii*) Strong vibrations of high intensity are recorded by ________.

(*viii*) Removal of people to safer places is called as ________.

(*ix*) Process of adding the strength to a weakend structure or damaged elements is called ________.

(*x*) ASR and MST stands for ________ and ________ respectively.

(*xi*) ________ is preferred where earthquake are more frequent.

(*xii*) Shapes like ________ are preferred for symmetry.

(*xiii*) Ductile structures show ________ damages.

(*xiv*) Diaphragm is another name of ________.

(*xv*) Zone V (having zone factor as 0.36) have ________ seismic intensity.

SECTION-B

2. Attempt any FIVE questions. **(5×6=30)**

(*i*) Define seismic waves. What are the different seismic waves?

(*ii*) Differentiate between magnitude and intensity of earthquake/

(*iii*) Why Seismic Codes are useful?

(*iv*) What is the importance of foundation in buildings?

(*v*) What is earthquake disaster mitigation?

(*vi*) What role do the slabs play during earthquake?

(*vii*) What is disaster management?

SECTION-C

3. Attempt any THREE questions. **(3×10=30)**

(*i*) Define earthquake. What are the various causes of earthquake?

(*ii*) What is seismic zone mapping? What is its importance?

(*iii*) Write short notes on following :

(a) Love waves, (b) Tsunamis

(*iv*) Define out-of plane failure. What are its main causes?

(*v*) Define retrofitting of structures. Why it is required?

❑❑❑

EARTHQUAKE RESISTANT BUILDING CONSTRUCTION

SIXTH EXAM/CIVIL/5132/6851/Nov., 2016

Duration : 3 Hrs. *M.Marks: 75*

SECTION-A

1. **Fill in the blanks :** **(10×1=10)**

(*a*) The waves which travel fastest and are first recorded are ________.

(*b*) Intensity of earthquake decrease with distance from the ________.

(*c*) The size of the opening in a building should be kept ________.

(*d*) The walls parallel to the earthquake motion are called ________.

(*e*) In earthquake resistant RCC structure minimum grade of concrete to be used should be ________.

(*f*) Removal of people to safer place is called ________.

(*g*) Column not having footing resting on ground is called ________.

(*h*) A fracture along which the blocks of rock slide is called ________.

(*i*) The major cause of earthquake is ________ causes.

(*j*) The revised seismic map divide india into ________ Seismic zones.

State True or False :

(*k*) Earthquake of minor intensity occurs every day.

(*l*) Masonry building are brittle structure.

(*m*) Short walls are more vulnerable to ground shaking than long walls.

(*n*) For earthquake resistant structure the weight should be large enough.

(*o*) Earthquake forces induced are proportional to mass of the structure.

SECTION-B

2. **Attempt any six questions.** **(6×5=30)**

(1) Give a shoft note on seismograph with diagram?

(2) Define the terms: (i) Focus, (*ii*) Epicenter

(3) What is out of plane failure?

(4) Explain splint and bandage technique of retrofitting?

(5) What is role of shear walls in RCC buildings?

(6) What is short column effect?

(7) What precautions should be taken in emergency phase i.e., during earthquake?

(8) Give a list of rescue equipment

(9) What do you understand by irregularities in building?

SECTION-C

3. Attempt any three questions. **(3×10=30)**

(*i*) Define seismic waves? What are different types of seismic waves? Explain important characteristics of P-waves?

(*ii*) Define magnitude and intensity of an earthquake? How will you related them?

(*iii*) Explain the technique and methods used in retrofitting?

(*iv*) Explain the main provisions made in IS 4326-1993.

(*v*) What is disaster management? Explain rehabilitation?

❑❑❑

References

1. IS : 456. Plain and Reinforced concrete-code of practice, Bureau of Indian Standard, New Delhi, 2000
2. IS : 1893, Criteria for Earthquake resistant Design of Structures-Part 1 : General Provisions and Buildings (Fifth Revision), Bureau of Indian Standard, New Delhi, 2002
3. IS : 13920, Ductile Detailing of Reinforced Concrete Structures Subjected to Seismic Force, Bureau of Indian Standards, New Delhi, 1993 (Reaffirmed -2003)
4. IS : 1893-1970 Indian Standard Criteria for Earthquake Resistant of Structures (Second Revision), Indian Standard Institute, New Delhi, 1971.
5. IS 4326 : 1993, Earthquake Resistant Design and Construction of Buildings – Code of Practice, Bureau of Indian Standard, New Delhi, (Reaffirmed-2005)
6. IS 1905 : 1987 Code of Practice for Structural use of Unreinforced Masonry, Bureau of Indian Standards, New Delhi, 1985.
7. SP 20 (S & T), Hand Book on Masonry Design and Construction, BIS, New Delhi, 1991
8. IS : 13827, 1993, Indian Standard Guidelines for Imposing Earthquake Resistance of Earthen Building, Bureau of Indian Standards, New Delhi, 1993.
9. IS : 13828, Improving Earthquake Resistance of Low Strength Masonry Buildings–Guidelines, Bureau of Indian Standards, New Delhi, 1993.
10. IS : 13935, Repair and Seismic Strengthening of Buildings Guidelines, Bureau of Indian Standard, New Delhi, 1993.
11. Agarwal, P and Shri Khande, M., "Earthquake Resistant Design of Structures", PHI, New Delhi, 2006.
12. Murty, C.V.R., "Earthquake Tips-Learning Earthquake Design and Construction", IIT Kanpur-BMTPC, New Delhi, 2005
13. "A Manual of Earthquake Resistant Non-engineered Construction, Published by Indian Society of Earthquake Technology, University of Roorkee, Roorkee, 1989

Index

S

T

U

V

Win Prizes

ATTENTION READERS

Win Prizes

We request for your frank assessment regarding various aspects of the book as given below :

EARTHQUAKE RESISTANT BUILDING CONSTRUCTION

By : NEELAM SHARMA

Please Fill up the Sheet in Neat and Clean Writing

(i) What topics of your syllabus, important for your examination point of view, are not covered in this book ?

....

....

....

....

(ii) What are the chapters wherein the treatment of the subject matter is not systematic or organised or updated ?

....

....

....

....

(iii) Have you come across misprints/mistakes/factual inaccuracies in the book ? Please specify the chapters and the page numbers.

....

....

....

....

(iv) What other books on the same subject have you found/heard better than the present book ? Please specify in terms of price and quality.

....

....

....

....

(v) Further suggestions and comments for the improvement of this book.

....

....

....

....

(PS : If need be please attach a separate sheet to write your views)

OTHER DETAILS

(*i*) Who recommended you this book ? (Please tick (✓) in the box near the option relevant to you.

☐ Teacher ☐ Friends ☐ Bookseller

....

....

....

....

(*ii*) Name of the recommending teacher, his designation and address:

....

....

....

....

(*iii*) Name and address of the bookseller you purchased the book from :

....

....

....

....

(*iv*) Name and address of your institution (Please mention the University or Board as the case may be)

....

....

....

....

(*v*) Your name and complete address :

....

....

....

....

(*vi*) Course you are enrolled in :

....

....

....

....

The best assessment will be awarded each month : The award will be in the form of our publications as decided by the Editorial Board.

Please mail the filled up coupon at your earliest to :

S.K.KATARIA & SONS®

4885/109, Prakash Mahal, Dr. Subhash Bhargav Lane,

Opposite Delhi Medical Association, Daryaganj, New Delhi–110002 (INDIA)

Phone: +91-11-23243489, 23269324, 43551243; *Telefax:* +91-11-23243489

e-mail: katariabooks@yahoo.com; katariabook@gmail.com

Website: www.skkatariaandsons.com; *For e-books:* www.katsonbooks.com